Yves Tourte

Genetically Modified Organisms

Transgenesis In Plants

Genetically Modified Organisms
Transgenesis in Plants

Genetically Modified Organisms
Transgenesis in Plants

Yves Tourte

Professor of Biology
University of Poitiers
France

Science Publishers, Inc.

Enfield (NH) Plymouth, UK

CIP data will be provided on request.

SCIENCE PUBLISHERS, INC.
Post Office Box 699
Enfield, New Hampshire 03748
United States of America

Internet site: http://www.scipub.net

sales@scipub.net (marketing department)
editor@scipub.net (editorial department)
info@scipub.net (for all other enquiries)

ISBN 1-57808-260-9

Published by arrangement with Dunod, Paris

This work has been published with the help of the French Ministere de la Culture- Centre national du livre.

© 2003, Copyright reserved

Translation of: ***Les OGM*** la transgenèse chez les plantes, Dunod, Paris, 2001.

French edition: © Dunod, Paris, 2001
ISBN 2 10 005279 9

Published by Science Publishers, Inc. Enfield, NH, USA
Printed in India.

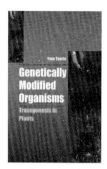

Preface

What some of us will remember most about the turn of this century and millennium is the atmosphere of crisis and polemics linking the most common of our concerns, food and its impact on our health, with the stakes of an ethics that touches our most profound convictions—the ownership of living things through the "patentability" of genes, the possible cloning of human beings, and the uncontrolled dispersal of transgenes. All these processes seem to be possible only through restriction of our individual liberties. This crisis is unusual in that it involves consumers concerned with their comfort, their health, and the future of their children as well as scientists who are aware of the stakes but do not know how to respond to the questioning or anguish of people confronted with new kinds of products.

Is this a true crisis in a society that has suddenly become aware of its moral and judicial destitution in the face of the potential power of scientists, who are themselves sometimes enslaved by industrialists or financiers? Or is it simply the kind of historical convulsion that civilization has felt at every enunciation of a new fundamental principle that challenges the achievements or convictions of its predecessors? Such convulsions occurred in response to the announcement of the rotation of the earth around the sun, universal gravitation, and the equivalence of space and time or of matter and energy. All of these ideas have greatly influenced our modern world view much more than they influenced the way of life of the contemporaries of these discoveries.

Crisis or convulsion? The answer is not easy and demands a certain distance that we do not yet have.

Still, would it not be advisable to take a look that, if not really objective, yet attempts to be at least prudent and honest, at the knowledge we now have access to, a perspective that is still only a preliminary approach to the problem of genetically modified organisms? GMOs are a subject about which everyone seems to agree on a "principle of caution", even though it encompasses widely divergent, even opposing, conceptions. Although, according to this principle, it is advisable to programme certain domains of research with an abundance of caution, we cannot justify an entirely wait-and-see attitude when the solutions to certain urgent problems, accessible only through this direction of research, may be within the grasp of scientists.

This has happened with certain solutions to problems of hunger and disease throughout the world. There is no progress without risk, but in what cases has there really been progress, and should we not be careful of increasing the risk by rashness or excessive prudence? The crisis, if there is one, has perhaps been best illustrated in the ambiguity of confident detractors of GMOs who simultaneously base an immense hope on gene therapy, when the two processes start strictly from the same principles and technologies.

We cannot eternally justify such an ambiguous position simply on the necessarily subjective consideration of the magnitude of the stakes involved.

YVES TOURTE

The author wishes to thank Carole Vauzelle,
Sandrine Champenoy-Flajoulot,
and Monique Tourte for their participation.

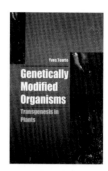

Contents

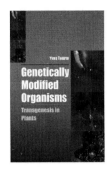

Chapter 1
Introduction

1.1. GENETICALLY MODIFIED ORGANISMS

The now familiar acronym GMO refers to living organisms that have been genetically modified. In the mind of the public, this term is generally associated with *commercial crops* that have been manipulated in a mysterious and disturbing manner. In fact, GMOs, strictly speaking, involve all living things—including animals, bacteria, or fungi—that have been genetically modified in the laboratory following the deliberate intervention of scientific researchers. The term is rather vague because it supposes that there are, in contrast and in an infinitely more natural way, genetically conforming organisms, which we know there are not. Moreover, scientists can easily demonstrate that there are not enough inhabitants on earth for two persons to have exactly the same genetic make-up. Even identical twins, although they are close to having the same genetic make-up, are differentiated by some errors, uncorrected, that occurred in the duplication of DNA during the single mitosis that separated the first two daughter cells from which they respectively developed. Every individual represents an original recomposition of the genetic make-up of its parents. There is thus no conformity except in the overall expression, which must guarantee the permanence and continuity of the species to which the parents and descendants belong, while ensuring the individual's constant ability to adapt to variations in environmental conditions.

Here we touch on the *two essential properties of the genome: stability and plasticity*. This is one of the most beautiful subjects of reflection for biologists, and one that is often offered to the sagacity of students writing biology examinations!

Animal or plant breeders frequently cross genetically very different varieties in the hope of obtaining a manifestation of *heterosis*. This activity permanently and profoundly modifies the organization of the genome. Even in individual plants belonging to populations that have multiplied by strict autogamy or vegetative propagation, there appears some variability, called somaclonal.

What is true at the individual level is more so on the cellular level. The zygote, which results from the fusion of two gametic cells that are highly differentiated but in very different ways, represents an original genomic composition. The zygote rapidly divides into daughter cells that then enter a process of differentiation accompanied by progressive loss of their totipotentiality. This process seems much more rapid and intensive in animals than in plants. Cell differentiation has long been investigated to find out whether it accompanies a modification that affects the genome in just its expression or in its structure. A partial answer was given in 1997 with the birth of the celebrated cloned sheep Dolly, a product of the union of a nucleus of a differentiated cell and cytoplasm of a female gamete. The answer was only partial because the premature ageing of the sheep could be interpreted as the expression of a modified cell.

Constant modification of a genome thus seems to be the rule and conformity the exception.

In reality, the description of GMOs is much narrower because their creation depends on the calculated intervention of scientists. One definition of GMO is *an organism whose genetic material has been modified in a way that is not possible by natural propagation and / or recombination* (French law no. 92.654, 13 July 1992). To create a GMO, the researcher must use a technology adapted to a programme or project designed to modify the organization and functioning of the genome of all the cells of the organism including, if possible, the reproductive cells. Genetic modification of only part of the cells results in chimeric organisms that cannot be considered true GMOs. The modification is due to the transfer of a gene taken from cells of another donor organism that must be introduced in the host organism: this is the *transgene*. This transgene must be expressed in its new environment, at least in a transitory manner and at best in a durable manner. In the latter case, it is often necessary for the transgene to be able to reproduce at the same time as the genome of the host cell; for this, it must integrate itself in the host cell. All these events must be verified. Geneticists set up a range of techniques that will be described in this work and that allow them to understand the future of this transgene and the consequences of its expression in the organism thus genetically modified. Many complementary techniques are available, but they are sometimes very time-consuming. We will see that one of the most delicate problems raised by GMOs at present involves their detection and their identification.

To be considered a GMO, an organism must be living. Thus, products derived from their metabolism or their cadavers are not strictly GMOs even though it is these products found in food that presently seem to pose a problem in the eyes of the public.

Before tackling the problems that GMOs pose in our time, it seems necessary to better understand the principles and methodologies used to create GMOs, particularly the major techniques of molecular biology that biotechnologists use, as well as the intended objectives of their creation. We will speak, later in this work, essentially of genetically modified plants, which are presently most affected by this technology. However, we must discuss bacteria and yeasts, valuable auxiliaries for the creation of vectors and identification of genes without which genetic transformation in plants would be impossible. We will also mention other categories of organisms—prokaryotes, viruses, fungi, and animals—that are affected by this technology and underline the differences in the technical approach and the results obtained. We will look at the achievements, attempts, and promises of these GMOs in large sectors of activity—agro-foods, medicine, and environment—before tackling the uncertainties, the stakes, and the investigations that their presence gives rise to in each of these sectors. We will study the regulatory, judicial, and bioethical responses of societies and nations with respect to GMOs.

GMOs are not the products of chance, as in the fantasies of mischievous science writers. They are the fruits of an evolution of biology in the second half of the 20th century. Let us first look at some highlights of their history.

1.2. A BRIEF HISTORY

GMOs are the product of a discipline of biology called "genetic engineering", itself an integral part of biotechnology. After having discovered the laws of genetics, it seemed natural that scientists would desire to control them. The first manifestation of this control can be considered the birth of genetic engineering. It is nevertheless very difficult to give genetic engineering a precise date of birth because mastery of the genome, in fact, took a long time to become a reality. Should genetic engineering be traced to the first successful transformation of a bacterium by the intermediary of DNA? If yes, the right answer would be 1944, when O.T. Avery successfully transformed strains of non-pathogenic (avirulent) *Pneumococcus* strains into pathogenic (virulent) strains by putting the first kind of bacteria in contact with DNA of the second kind *in vitro*. Only an overall control was achieved, on the population or colony level, and the interpretation was only approximate and incomplete.

It seems more reasonable to consider that genetic engineering was born in the early 1970s, with the discovery of remarkable tools such as restriction enzymes, a sort of molecular scissors to cut DNA with, and the perfection of the first cloning vectors. Paul Berg presented in 1972 the first studies on cloning, during which he used the first restriction enzyme extracted from the bacterium *E. coli*, the endonuclease EcoRI, to linearize DNA of a virus (SV 40) as well as to cleave the DNA of a bacterial plasmid carrying a gene for galactose metabolism. Normally, bacteria are equipped with these enzymes to defend themselves against parasitic bacteriophages. Berg thus succeeded in creating the first plasmid vector capable of modifying the genome of a host cell (Inset 1.1).

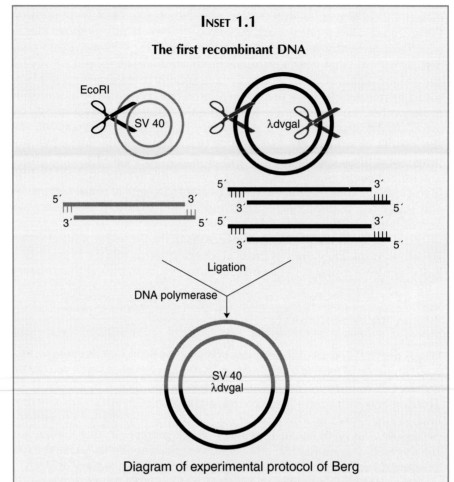

INSET 1.1

The first recombinant DNA

Diagram of experimental protocol of Berg

Paul Berg was the first to carry out a manipulation of DNA cloning. He used the first endonuclease discovered, EcoRI, to linearize the virus SV 40 and to clone a part of the DNA of a bacterial plasmid. This plasmid normally allows the bacteria to utilize galactose (λdvgal). EcoRI can be used to obtain fragments with sticky ends (which was unknown to Berg, who took the trouble of adding complementary sequences designed to link the fragments to one another, whereas the enzyme spontaneously generated such sequences). He thus obtained the first recombinant DNA ligating the bacterial DNA and the viral DNA.

The 1970s were a particularly fertile period for "molecular biology", in its fundamental aspects as well as for some applications. There was an extraordinary development of research on restriction enzymes extracted from a large number of species of bacteria and capable of cleaving the DNA molecule at highly precise sequences. Some enzymes make straight cleavages whereas others, which are more numerous, have the peculiarity

INSET 1.2

Restriction endonucleases

The first diagram shows a staggered cut on the two strands, as with EcoR1; the ends of fragments can spontaneously ligate again.

The second diagram shows a straight cut, as with the enzyme Hae III. The ligation is much more difficult to achieve, impossible in some cases. The table gives some examples of endonucleases and the types of cuts resulting from their use. Today we know of over 100 different endonucleases.

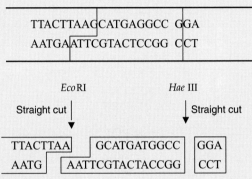

Name of enzyme	Extract of :	Recognized sequence	Type of cut
1 *Eco* RI	*E. coli*	...G‖AATTC... ...CTTAA‖G...	5′ end
2 *Taq* I	*Thermus aquaticus*	...T‖CGA... ...AGC‖T...	5′ end
3 *Cfo* I	*Clostridium formicoaceticum*	...GCG‖C... ...C‖GCG...	3′ end
2 *Hae* III	*Haemophilus aegytius*	...GG‖CC... ...CC‖GG...	Straight cut
3 *Dra* I	*Deinococcus radiophilus*	..TT T‖AAA ...AAA‖TTT	Straight cut
4 *Eco* R V	*E. coli*	..GAT‖ATC... ..CTA‖TAG...	Straight cut
6 *Mbo* II	*Moraxella bovis*	...AGAN‖... ...TCTN‖..	N = purine or pyrimidine
7 *Hin* fI	*Haemophilus influenza* R$_f$	...G‖ANTC... ...CTNA‖G...	N = purine or pyrimidine

of generating sticky ends, which tend to reassociate spontaneously. These enzymes can thus be used to excise fragments from a DNA molecule and replace them by other fragments cleaved by the same enzyme and carrying other genes. This is the basic principle of genetic modification. Once the new fragments are inserted, they must be able to be conveniently exploited by the host cell and transmitted, during cell division, to two daughter cells (Inset 1.2).

The progress in this field was so rapid that the first private organization based on controlled exploitation of the genome, Genentech, was created in 1976 in the United States by H. Boyer. This first society of biotechnology produced human somatostatin by a genetically modified bacterium (or "reprogrammed bacterium", as some call it).

By the end of the 1970s, the bacterial plasmid, for example plasmid pUC 18 (Inset 1.3), which was equipped with its source of replication as well as a polylinker or cloning site and could be opened, closed, cut, completed, and transformed, became a familiar object in all molecular biology laboratories.

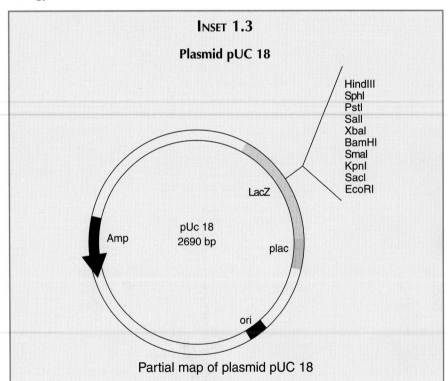

Inset 1.3

Plasmid pUC 18

Partial map of plasmid pUC 18

This bacterial plasmid of close to 2700 base pairs, like all plasmids, has a source of replication that ensures that is it is copied, a gene for resistance to ampicillin (amp), the gene lacZ with its promoter (plac) and a multiple site of cloning that allows the insertion of exogenous DNA sequences by means of specific restriction enzymes.

The transformation of eukaryote cells was achieved in the early 1980s, with the perfection of transfection techniques. It can be said that at this time no biological group refrained from attempts at genetic transformation. In the plant kingdom, genetic transformation of cells became part of the perspective of a new agronomy looking beyond its usual agro-food prospect toward development at the agro-industrial level. As for animal or human cells, their transformation would interest the medical field first of all, i.e., pharmaceuticals and cosmetics. We will see that the development of genetic engineering in these two key sectors of the economy—agro-food and medicine—would be perceived in very different ways by the public.

A third major sector of the economy, that of the environment and all that follows from it for our framework of life, was promptly involved in genetic engineering. Here also, not without polemics and confrontations.

Inset 1.4

PCR: Polymerase chain reaction

The thermocycler is an apparatus that allows the alternation of periods of high heat designed to denature DNA, which must then become single-stranded, and cooling periods during which a heat-stable DNA polymerase (extracted from a thermophilic bacterium) synthesizes complementary nucleotides of DNA strands separated during the preceding cycle. The thermocycler does this with very low thermic inertia and for perfectly determined time periods. The polymerization necessitates specific primers from precise regions of DNA that are to be amplified: A1 and A2E. The original chains (but also the copies, which become exponentially more numerous) are thus copied again at each cycle of partial cooling. About 30 cycles are required to obtain quantities of DNA that can be manipulated.

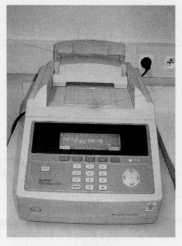

Thermocycler

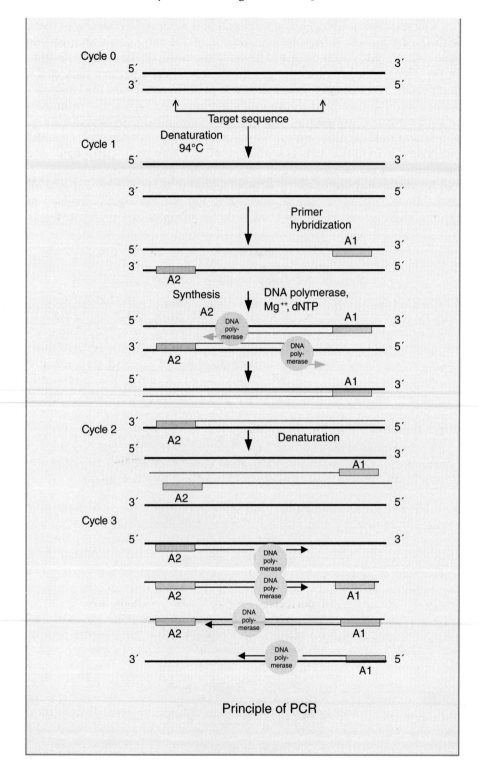

Principle of PCR

An original and critical invention appeared in the 1980s that contributed greatly to the development of genetic engineering, enabling easy and reliable *in vitro* amplification of DNA: polymerase chain reaction or PCR (Inset 1.4). This technique is now routinely used in laboratories. Industries have manufactured small robots called "thermocyclers", convenient, reliable, and affordable, that made it possible to automate the reaction; these robots are now part of the inventory of any molecular biology laboratory.

Why, after having mastered genetic transformation of prokaryotes, did scientists take several years to perfect the techniques of genetic transformation of eukaryote cells? The answer lies in the varying complexity of these two types of cellular organization. In both cases, however, the DNA molecule to be transferred must cross the plasma membrane. It has been known since the 1960s how to make the membrane of a bacterium permeable by using anions to neutralize the electric charges of the DNA molecule as well as of the membrane.

Subsequently, it was discovered how to obtain identical results with equipment called electroporators, also capable of neutralizing the electric charges by a brief but relatively intense electric shock. We can also mask the electric charges rather than neutralize them by wrapping the living matter in a plastic film such as polyethylene-glycol in a concentrated solution. Some of these techniques were responsible for successful transformation of eukaryote cells. The parameters of electroporators could be modified to create equipment for electrofusion, generally by means of specialized additional modules. Successes were regularly recorded, but the yield was always discouraging low. Also, the access to transformation of eukaryote cells was much greater by "indirect" methods consisting of first confiding the material to be transferred to biological vectors—viruses and certain bacteria—capable of invading the tissues of the host and inserting the foreign DNA in the genome of some of the host cells. Such indirect methods were developed particularly for the transformation of plant cells by the intermediary of bacteria of soil origin that could parasitize plants. This method was initiated in Belgium and today has become routine throughout the world. For the transformation of animal cells and for a possible application to gene therapy, it is the viruses that are generally favoured. The development of these indirect methods, however, did not prevent the pursuit of perfection in the direct methods already cited. Another direct method, developed in 1987, is the technique called "biolistics", which we will describe in a later part of this work.

The indirect techniques took up the strategies already naturally used by parasites to domesticate their host cells. For many years now, the tactic used by bacteriophages to modify the genome of their bacteria-hosts has been known. Also, it has been known that the bacteriophages can end up with a genome modified by the addition of a small sequence of DNA taken from the genome of the bacterium. This sequence could itself be transferred into the genome of a second bacterium parasitized by the bacteriophage.

This is the phenomenon of transduction. The scientist therefore seeks to use and control this capacity of the bacteriophage to transfer selected information from one bacterium to another (Inset 1.5).

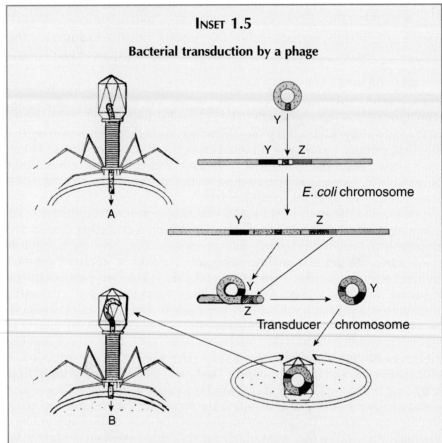

INSET 1.5

Bacterial transduction by a phage

Principle of bacterial transduction by a phage

Phages (or bacteriophages) are parasitic viruses of certain bacteria. Transduction is the transfer of a gene belonging to one bacterium (A) into a second bacterium (B) by the intermediary of a phage that successively parasitizes the two bacteria by momentarily integrating its genome with theirs. Sometimes, albeit rarely, the phage, after being integrated in the genome of bacterium A, reindividualizes itself by carrying in its own genome a small fragment of the genome of the host bacterium, which it will transmit into bacterium B when it parasitizes B. Bacterium B thus acquires a new property due to the gene of bacterium A that the phage has transmitted to it. This is an operation of natural genetic transformation. At the molecular level, the transfer necessitates the presence of a homologous sequence between the DNA of a phage and a region of bacterial chromosome, a region that allows the integration of the phage DNA into the bacterial chromosome. When the virus organizes itself to lyse the bacterium, it forms an excision loop. This excision may be normal and will isolate the DNA of the phage in its primitive organization but it may also be abnormal and thus lead to a small fragment bordering the bacterial DNA in the phage DNA. It is this small fragment that will be transferred to bacterium B when the virus begins to parasitize.

Taking into account the considerable number of viruses that can parasitize animal cells, transgenesis in the animal kingdom is essentially a matter of choosing the right virus. This choice takes into account its pathogenic or non-pathogenic character as well as the specificity of target cells.

In plants also, viruses were used, particularly those belonging to the geminivirus category, but they were quickly abandoned in favour of *Agrobacterium*. The damage this bacterium causes in the crown of a large number of cultivated plants has long been known, although the actual mechanism by which it does so was not known. It was only during the 1980s that the proliferation of plant cells that characterizes this disease, crown gall disease, was shown to be due to natural transfer of a fragment of a large plasmid that contains the agrobacterium into the genetic make-up of the parasitized cell. The transfer is followed by integration of the fragment into the genome of the host cell (Inset 1.6).

INSET 1.6

Crown gall disease

Tumour characteristic of the disease

Crown gall disease is manifested by the formation of a tumour at the crown of the plant, i.e., at the region of transition between the root and the stem, a region that coincides with the soil surface. The disease affects mostly dicotyledons (cabbage, for example) and often leads to

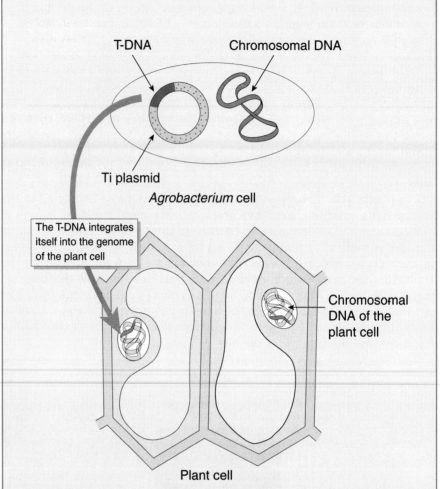

T-DNA

Chromosomal DNA

Ti plasmid

Agrobacterium cell

The T-DNA integrates itself into the genome of the plant cell

Chromosomal DNA of the plant cell

Plant cell

Mechanism of action of bacteria responsible for crown gall disease

the death of the plant. It occurs because of injuries due to shocks or climatic accidents (such as frost). It is observed that the tumour is invaded by a colony of rod-shaped bacteria, aerobic and gram negative. Under the influence of these bacteria, the plant secretes unusual substances, opines, that favour the proliferation of bacteria. Research in molecular biology has shown that these bacteria contain large plasmids, one part of which, T-DNA, is transferred into the nucleus of the peripheral cells of the plant and integrated in the genome of these cells. It is this transfer DNA that is responsible for the synthesis of opines. It is a natural genetic transformation in the plant cell. Knowledge of the mechanics of this disease is essential for the perfection of modalities of directed genetic transfers.

The system could not, however, be used in this state, considering the drastic effects of the bacterial DNA on the development of the plant. It was necessary to remove the oncogenes carried by this plasmid without altering its capacities of expression and integration. This was successfully done by Van Montagu and colleagues at the University of Gand, in the early 1980s. These researchers not only eliminated the oncogenes but replaced them, in carefully monitored conditions, with other genes whose transfer and expression were achieved. Thus, "disarmed" *Agrobacterium* was made available and plant transgenesis could develop further.

In the early 1980s, there did not seem to be any further technological obstacles to the transformation of all living things, no matter what groups were considered. Still, for these techniques to emerge from the basic research laboratories, there had to be a real need and, in consequence, a real demand by agents of the traditional economic sectors of agriculture and medicine.

In agronomy, the conventional techniques of plant breeding seemed to have achieved a great many objectives but were always considered costly because of the very long time required (10 to 12 years, on average) to develop a new variety that performs better than its predecessors. Certainly, in alternating the cultivation of successive generations from one hemisphere to another to benefit from two favourable seasons each year, some precious years can be saved. This is, and always has been, true of the technique of hybrid crosses, followed by backcrosses in such a way as to transfer into a plant that already performs well only the gene or the few genes capable of correcting a possible weakness. For example, we often introduce a disease resistance coming from a wild species into a cultivated variety, without losing the agronomic quality possessed by the cultivated variety. The dream of the breeder has always been to be able to extract just one gene from a species possessing a useful property and directly insert it into the variety of the breeder's choice. Genetic engineering has proved that it is the answer to this dream.

Genetic engineering, of course, has very quickly reached its limits, because there are only two ways, in principle, to obtain an organism in which all the cells are modified: either to transform all the cells of the plant or to take a transformed cell in such a condition that it can divide and develop into an entire organism, i.e., a cell endowed with the property of regeneration. The first solution very quickly proved to be impossible because, despite its performance, *Agrobacterium* transforms only a limited number of cells. To obtain genetically modified plants, therefore, it is necessary to start with proliferation of cells derived from transformed cells. It is here that plants have the advantage of a capacity of ontogenesis and organogenesis in practically all the cells, unlike animals, in which only the egg cell and its immediate derivatives are capable of developing completely. In plants, there is, of course, the zygote resulting from fertilization but there are also stages of microspore and macrospore and practically all the differentiated cells, a few of which need to be placed in appropriate culture conditions. This is the same principle of regeneration, a property particularly developed in plants because the processes seem to be, in this respect, in principle not accessible to animal cells. The domain of totipotent cells has plenty of surprises in store.

Plants proved themselves to be well adapted to techniques of transgenesis and the first genetically modified plants, tobacco, appeared in 1983. There was no specificity of transforming genes and it was very quickly proved that the genes of bacteria and yeast could be used as well as those of plants or animals. The first plants secreting an insecticide, after insertion of a gene for a bacterial toxin, were created in 1985. This was a tobacco that had received the gene for toxin production from the bacterium *Bacillus thuringiensis*. Today, in France and other countries of western Europe, entire fields of maize protected in this manner against attack by a species of caterpillar are cultivated after authorization by competent administrative bodies. Within a few years of the development of transgenic tobacco, commercial transgenic crops of soybean, maize, rice, and cotton were grown in some of the American states that replaced conventional plant varieties. Some were resistant to parasites, others to herbicides, and still others had a double resistance.

Moreover, for the past 30 years, crop surpluses pushed agriculture towards products designed for industrial purposes. The objectives were not the same and the farmer was confronted with new criteria of quantity, quality, regularity of production, chemical or technical specificity, and price, which always was quite low. In Europe, the common agricultural policy divided farm land into two categories: land for food crops and land that could be left fallow or planted with crops having industrial uses, including energy generation. We may find that this purpose will increase during the new century. When genetically modified plants may be consumed as food, the reservations are greater than when they are oriented toward plasturgy, materials for our environment, or our gas tank.

Curiously, there is one field in which the presence of GMOs has been not only tolerated but welcomed. This is the field of medicine, particularly pharmaceuticals. All types of GMOs are found in this field: bacteria to synthesize insulin or growth hormones, tobacco or lucerne reprogrammed to synthesize haemoglobin, and cultures of animal cells designed to synthesize recombinant proteins with therapeutic effects.

The perspective that the public is most likely to accept is that which, curiously, may pose the greatest concern and ethical reservations on some points. This concerns the genetic transformation of human cells for the purpose of repair or correction, a method still called "gene therapy". Every day we come across implacable enemies of genetic engineering who throw themselves energetically into national campaigns on behalf of victims of myopathy or mucoviscidosis. The technological process is identical in both cases, even though the objectives are different. GMOs and the rational are not yet in total harmony in the minds of many of our contemporaries.

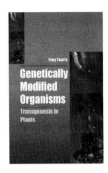

Chapter **2**

Techniques of Creating GMOs

For a GMO to be created, there must be an organism that donates the gene and a host organism that will become the GMO to the extent that the gene it receives is present and functional in all its cells. If only some cells are transformed, we have obtained only a chimeric organism with a much more limited use.

The major stages in the creation of a GMO are as follows:

- isolation of a functional gene or different parts that must serve to construct a functional gene;
- construction of this functional gene and its association with a selectable marker gene;
- insertion of this gene in a transfer vector and then transfer of the gene from donor to host;
- control of the presence and proper functioning of the transferred gene (or transgene) in the host;
- assessment of the stability, harmlessness, and risks of dispersal of the transgene.

We will thus study successively the stages in the creation of a GMO by giving, for each stage, the various techniques possible depending on the donor and host organisms.

2.1. ISOLATION OF A FUNCTIONAL GENE

2.1.1. Genome and genetic information

Living things, in each of their cells, carry information in one or several DNA molecules (Inset 2.1)

INSET 2.1

DNA molecule and replication

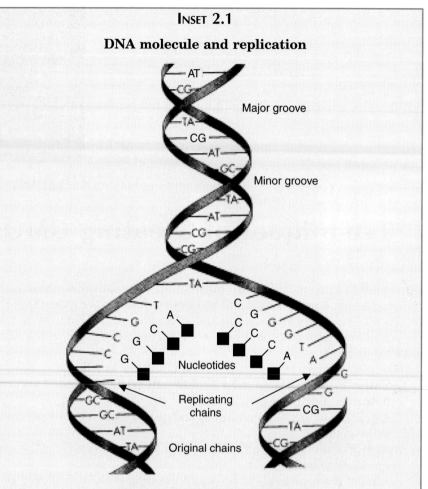

Replication of a DNA molecule

The genetic material of each cell is carried by long molecules of deoxyribonucleic acid or DNA. This molecule is well known since the studies of Watson and Crick and is structured in the form of a double helix (resembling a spiral staircase). Its "banisters" are formed by deoxyriboses linked by molecules of phosphoric acid and its "steps" are made up of four types of purine and pyrimidine bases: adenine (A), guanine (G), thymidine (T), and cytosine (C). T links only with A, and G links only with C. The succession of these bases all along the molecule constitutes the characteristic sequence of the genetic information, strictly speaking, of the region of this molecule being studied. The unit of information is the triplet, i.e., the succession of three bases. The triplet codes for the fixation of an amino acid during protein synthesis. One DNA molecule corresponds to one chromosome. The cells of rye, for example, contain 14 DNA molecules assembled in pairs (2n = 14). The reproductive cells or gametes contain only 7 entirely different molecules. Human cells contain 46 DNA molecules corresponding to 46 chromosomes.

The DNA molecule replicates in a semi-conservative way. Each "step" of the spiral can divide into two halves, each half remaining linked to its "banister". Each "half-staircase" can, by means of different molecules including specific polymerases, reconstitute the other half. This is the semi-conservative replication of DNA, and only this molecule (apart from prions, perhaps) possesses such a property.

In the prokaryotes, this information is represented by a single circular molecule, protected and located within the non-compartmentalized cell. In the eukaryotes, the information is carried by several molecules, in fact to the extent that there are chromosomes visible during cell division. These molecules, associated with histone proteins, are enclosed in a compartment, the nucleus (Inset 2.2).

INSET 2.2

Comparison of organization of genomes in the prokaryote and eukaryote cell

The genetic information in the prokaryote cell (e.g., bacteria) is carried partly by a long, circular molecule of DNA, closely resembling a skein of wool, which constitutes the single chromosome, and partly by small, circular molecules dispersed in the cytoplasm called plasmids. The chromosome carries some thousands of genes, while the plasmids carry only some tens, often genes for antibiotic resistance.

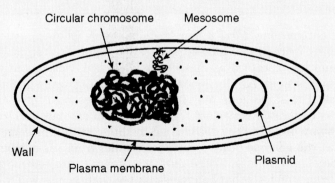

Genome of the prokaryote cell

The genetic information in the eukaryote cell is much more complex and has a compartmentalized support. In the nucleus, the genome is represented by several chromosomes (often several tens) that correspond to linear DNA molecules. These molecules ligate with proteins called histones to constitute elementary structures—the

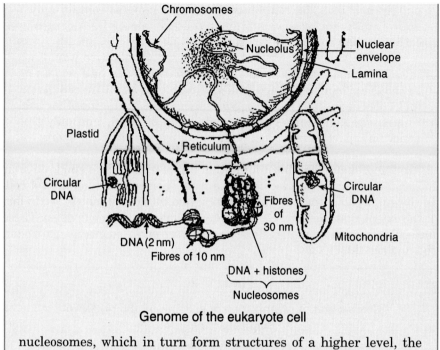

Genome of the eukaryote cell

nucleosomes, which in turn form structures of a higher level, the nucleosomal fibre of 30 nm. The nucleosomal fibre more or less condenses itself to form chromatin and, at its maximum level of condensation, the chromosome visible under microscope during nuclear division or karyokinesis. Genetic information is also present in the cytoplasm, which is found in the mitochondria and plastids and the organization of which is quite similar to that of bacteria.

Each chromosome carries thousands of genes, following one another, separated by more or less long regions called non-coding regions because they cannot be expressed as messenger RNA and, consequently, cannot code for any protein. All the genes together constitute the genome of the individual. The same gene may be present in several copies and these copies may be located very close together (in tandem) but also sometimes very far away because they are carried by different chromosomes. Let us recall that the genome of an individual results from the combination of two genomes of gametes. These gametes, having resulted more or less recently from a meiosis that has redistributed their genes, will, by uniting at random, generate an original genomic composition for each new individual.

2.1.2. Genes

Nearly all genes are constructed on the same principle. Each gene corresponds to a linear sequence of DNA of some hundreds to a few thousands of base pairs and generally comprises three basic parts: the promoter, the coding sequence, and the terminator. There are also regulatory sequences, but they are not necessarily continuous (Inset 2.3).

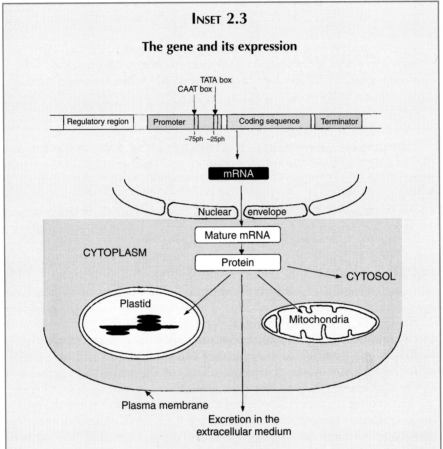

INSET 2.3

The gene and its expression

The gene and its expression

With respect to its organization, the gene is a DNA segment on which it is possible to distinguish three fundamental parts: the promoter, which initiates and controls the transcription of the next coding sequence, the coding sequence itself, whose originality is found in the protein synthesized, and then the terminator, which stops the process of transcription.

The promoter contains at least two "boxes" of regulation: the CAAT box and the TATA box. Its functioning is regulated by signals that it receives either directly or through an entire cascade of reactions from outside the cell.

The coding sequence is the transcribed part in a primary messenger RNA (or pre-messenger) within the nucleus, a messenger that will, in most cases, undergo maturation during its transfer in the cytoplasm (excision of introns and splicing of exons). The mature messenger will be expressed in a protein that often possesses a provisional mailing sequence indicating to it its final destination: maintenance in the same place, incorporation in a membrane, migration into an organelle, or rejection by simple secretion.

The secondary role of the terminator is often to fix a poly(A) tail, a distinctive sign of messenger RNA that ribosomal and transfer RNA do not possess. These two last types of RNA are involved in the translation of the messenger into a protein characteristic of the gene sequence.

The promoter, some hundreds of base pairs long, controls downstream the reading and thus the transcription or non-transcription of the coding sequence. It is recognized by the presence of some characteristic sequences having positions nearly identical in all the promoters of genes in eukaryotes: a CAAT box and a TATA box. The CAAT box is a portion of the promoter sequence located around 75 base pairs from its terminal part that plays a role in the fixation of RNA polymerase. The TATA box is a short sequence, rich in A-T linkages, located around 25 base pairs from the end of the promoter, on which transcription factors fix. The functioning of the promoter is regulated by factors internal to the nucleus or arising from the cytoplasm. It may even be induced by factors external to the organism (variations in physical or chemical conditions, biological stress). In such cases we speak of inducible promoters. There are cases in which the promoter is insensitive to these factors and functions in a simple regime under any circumstances. These are called strong constitutive promoters, and promoters of viral origin are an example.

The coding sequence, also called the reading framework, is the part of the gene that is transcribed in messenger RNA by RNA polymerase. The messenger will then itself be expressed in protein. The size of this sequence varies according to the genes, but it can be considered to be commonly between 1100 and 1500 base pairs. The sequence begins with a start codon, often an ATG, and ends with a stop codon (UAG, UGA, or UAA). The messenger may be directly expressed in a protein, but most often it undergoes a maturation that could comprise terminal or intercalary excisions. The excised parts correspond to introns and the conserved parts correspond to exons. The messenger RNA formed at maturity during its transfer into the cytoplasm through the nuclear pores is thus shorter than the native messenger RNA. Once protein synthesis from RNA is completed, the protein may either remain in place in the cell cytoplasm or migrate across the network of internal membranes (reticulum, Golgi bodies) towards its destination. In this last case, it has a signal sequence (sometimes two) that allows it to cross membranes and be correctly addressed. This sequence is subsequently excised. The protein also undergoes a maturation characterized by particular withdrawals and by the fixation of sugars (glycosylation of the protein). Sometimes there is a structure protein that is involved in the organization of membranes or organelles but most often an enzymatic protein that intervenes in the catalytic metabolism of the cell.

The terminator is a short sequence that causes the termination of the transcription. It is not rare for it to possess a polyadenylation sequence. This poly(A) tail characterizes the messenger RNA and differentiates them from other RNA (transfer RNA and ribosomal RNA), which collaborate with the messengers in protein synthesis.

The regulatory sequences are sometimes difficult to indicate because they can be located in distant parts of the genome. They function in a network and may influence the functioning of several genes. They serve as linkages between information from outside the cell and the promoters of genes. The information may pass directly into the nucleus or transit through an entire cascade of cytoplasmic reactions that trigger the involvement of secondary messengers such as proteins of the G-protein type. These regulatory sequences are not strictly part of the gene.

Thus, to be functional, the gene must have the three elements described in proper sequence. Artificial genes can be created by assembling the three types of elements taken from various genes, as long as the order and the proper sequences are respected. Such genes are chimeric and can be perfectly functional.

The information thus always descends from the DNA to the protein via the RNA. In the laboratory, we can "redirect" the protein towards the DNA, using rather complex protocols. Biochemists can decipher the chain of amino acids that constitute a protein. Starting from that protein sequence, it is possible to synthesize the corresponding RNA in vitro. By using this RNA along with a particular enzyme isolated from certain RNA viruses, inverse or reverse transcriptase, it is possible to synthesize complementary DNA or cDNA. This DNA contains only information corresponding to the coding sequence obtained from mature RNA. However, preceded by a promoter and followed by a terminator, the sequence in its context of chimeric gene could be the source of a functional protein identical to the original protein.

2.1.3. DNA libraries

The creation of a GMO always begins with the sequencing and isolation of the gene that one wishes to insert. To procure this gene, it is necessary to look for an organism in which the operation has the greatest chance of being successful and easy. The first operation consists of constituting a gene library. There are two types of gene libraries: genomic DNA libraries and libraries of complementary DNA or cDNA.

a) Genomic DNA libraries

To put together a genomic DNA library, a sufficient quantity of organisms or a homogeneous population of cells must be available from which all the DNA can be extracted. There are various methods of DNA extraction and purification depending on its origin or its location. For example, to extract DNA from bacterial plasmids, we can follow the protocol given in Inset 2.4.

INSET 2.4

Extraction of plasmid DNA

To obtain plasmid DNA of bacteria, we start from a known quantity of a bacterial culture that is centrifuged for 2 min. at 5000 g. The residue is put in alkaline lysis solution in the cold that allows the rupture of cell membranes and consequently the destruction of cells. The cell debris, the ruptured membranes, and the genomic DNA that is linked to these membranes are eliminated by a second centrifugation. A solution of phenol, chloroform, and isoamyl alcohol is used to eliminate the proteins. Two more rinses are done in the same conditions to bring the contents to high purity. The upper phase of the contents of the tube obtained after the second centrifugation is eliminated, while the lower phase, which contains the DNA and RNA, is precipitated by the addition of isopropanol. A treatment with RNase at ambient temperature allows us to conserve only the plasmid DNA in the tube. There are many variants of these techniques and the choice depends on the degree of purity desired. (*See diagram on page 23*)

The DNA is subsequently treated by a restriction enzyme that results in the formation of sticky ends. Likewise, the plasmid DNA extracted from a clone of bacteria carrying a gene for antibiotic resistance is treated in the same way and by the same enzyme. This treatment opens up the plasmid by generating identical sticky ends. The two DNA are mixed together and the plasmid DNA and genomic DNA fragments meet at random, leading to the formation of new plasmids that are slightly larger and carry fragments of genomic DNA. Some of these fragments are large enough to contain one or several genes. Thus, plasmids carrying a set of genes coming from a donor organism are made available. These plasmids are then put in contact with bacteria and, following a thermal or electrical shock, penetrate the bacterial cells. The bacteria are then cultured in the presence of an antibiotic and only those cells that have incorporated the plasmids can survive. These bacteria carry both the gene that gives them resistance to the antibiotic and the genome fragments of the donor organism. If we work on sufficient quantities of plasmid DNA, we can make available a sufficient quantity of bacteria for which almost all the genes of

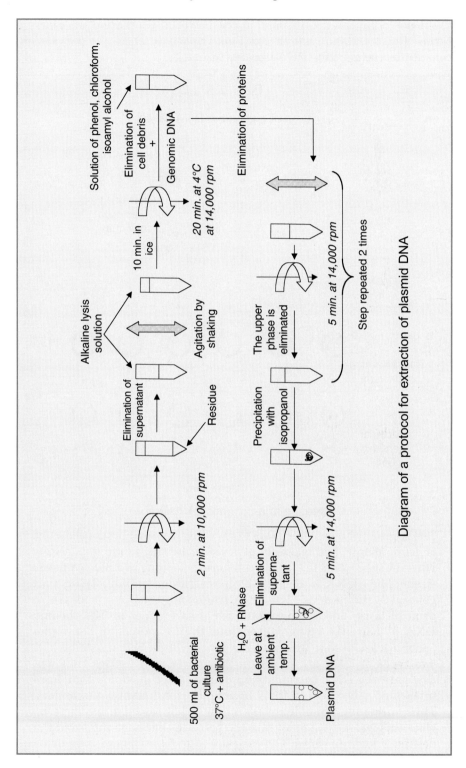

Diagram of a protocol for extraction of plasmid DNA

the donor organism are present in the bacterial population. In concrete terms, the petri dish, which contains the entire bacterial population, constitutes the genomic DNA library (Inset 2.5).

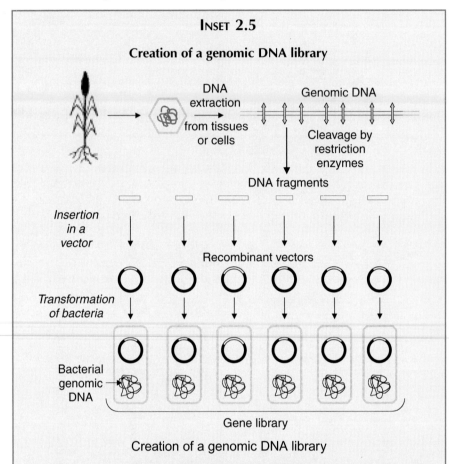

INSET 2.5

Creation of a genomic DNA library

Creation of a genomic DNA library

The creation of a genomic DNA library involves eukaryote cells, cells of young plants, and bacterial plasmid vector. As a first step, all the DNA must be extracted from a cell mass belonging to a homogeneous tissue. This DNA is cleaved by a restriction endonuclease that generates sticky ends. The DNA fragments thus obtained are in principle large enough to contain one or several genes. The plasmid DNA of bacteria is extracted and then cleaved by the same endonuclease used to fragment the genome of the original plant. The two types of DNA are mixed in the presence of ligases that will associate the DNA fragments and reconstruct plasmids that have integrated the different DNA fragments of eukaryotes at random. Thus, we obtain recombinant DNA that will be cloned in the bacteria. Once this is done, we must find the bacteria that have inherited the desired eukaryote genes. The easiest way to find the gene is by using a complementary probe that will be capable of hybridizing with it. If no such probe is available, for example when the gene has never yet been identified in any other organism, we must proceed to an operation of functional complementation. This consists of using a mutant, for example, a baker's yeast that is mutant for the desired function, and re-establishing the wild phenotype by incorporating the DNA fragments at random until one of the yeast cells gives rise to a colony growing in a minimal environment.

b) cDNA libraries

Genomic DNA libraries are sometimes difficult to use. cDNA libraries are preferred, although they are less complete, because they are easier to manage. To put together such libraries we start by extracting all the messenger RNA in an organism. These messengers involve only genes expressed at the time of extraction, so we must also choose a sequence of development of the organism corresponding to the moment at which the largest number of genes will be expressed. In the plant variety used for creating the library, we take young embryos at the time the first organs develop. Obviously, if we are looking for a gene involved in the flowering process, we must start with a much later stage, for example the time of emergence of the floral button; such genes are "silent" in the young embryo. It will be easy, from these messengers, to trigger the functioning of reverse transcriptase to obtain the corresponding cDNA. We proceed by introducing the DNA fragments into the plasmids of a bacterial colony that will ensure their multiplication. Here also, concretely, the library is represented in the form of a petri dish containing many bacterial clones, each carrying a cDNA fragment corresponding to one of the coding sequences of the genes (Inset 2.6).

INSET 2.6

Creation of a cDNA library

When looking for a gene, researchers often use a library of complementary DNA or cDNA. They can only, in fact, find here the coding sequences lacking introns. The construction of this type of library requires the extraction of messenger RNA of cells that are fully active, such as young seedlings or young plantlets at the beginning of their development. This is considered a period of their life during which these organisms use a maximum number of genes to carry out their metabolism. Let us note, nevertheless, that this stage will not give us genes linked to reproductive processes such as genes of floral organ development. For that purpose it is advisable to use plants at the floral initiation stage.

The extracted mRNA in the presence of reverse transciptase enzyme generate single-stranded complementary DNA molecules. These molecules then duplicate to give double-stranded DNA corresponding to coding sequences of genes that have given rise to messengers. These DNA are then incorporated in the bacterial plasmids by means of ligases. The plasmids possess an antiobotic resistance marker and only the bacteria that incorporate these plasmids will survive in the selection medium. These plasmids are a hybrid of cDNA and plasmid DNA and after multiplication will constitute the DNA library.

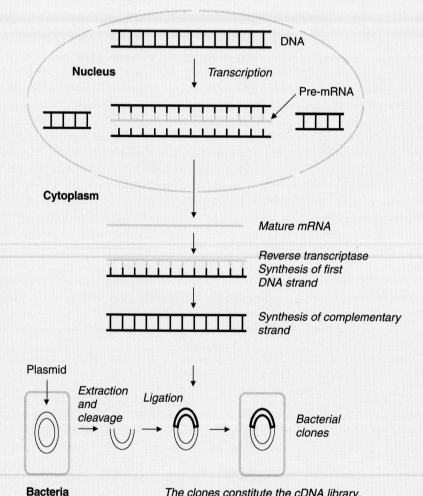

Creation of a cDNA library

The gene is then identified as in the case of a genomic DNA library either by means of a probe if one is available or by functional complementation of spontaneous yeast mutant or mutants created by gene disruption.

2.1.4. Searching for a gene in a library

Searching for a type of gene in a library is a delicate process that can be undertaken in many ways. In a small number of cases, the gene is identified in another living group, even a distant group from the phylogenetic point of view. A copy, even a partial copy, of this gene could serve as a probe in the search for its homologue in the library. In most cases, no other candidate is available and, in fact, only the function of the protein that is coded by this gene is known. In this case, the technique of functional complementation is used most often. A mutant colony must be available (colonies of bacteria, yeast cells, cultured cells), i.e., organisms incapable of realizing one function. These organisms can only survive in culture conditions in an appropriate nutrient medium that compensates for the defect or fault of the colony. Cultured on a minimal medium, they will soon die and only those organisms can survive that have recovered the lost function. This is the very basis of the method. The mutant colony is put in the presence of plasmids extracted from the library. An appropriate treatment of membrane permeability allows the mutants to recover plasmids at random and, in this lot, there may be some mutants likely to recover a plasmid vector of the gene that is capable of restoring the desired function. These mutants thus regain their wild character and are the only ones that contain the desired gene in their plasmid. From analysis of this plasmid DNA, after multiplication of individuals, the gene can be isolated and sequenced. The isolated gene can be introduced into a transfer vector to achieve the genetic transformation of the host.

A researcher who does not have a mutant available can induce one either by the action of mutagenic agents or by insertion of small DNA fragments in the genome at random. Some of the organisms lose the desired function and can thus survive only when cultured in a supplemented medium. There is genetic disruption and this induced mutant is then treated in the same way as the mutant described earlier. This is also one of the roles of transgenesis (Inset 2.7).

INSET 2.7

Gene disruption by random DNA insertion

In some organisms, such as baker's yeast (*Saccharomyces cerevisiae*), researchers make use of the thousands of mutants affecting a considerable number of essential metabolic functions. These yeasts can nevertheless survive on supplementary medium and then be stored at −80°C.

They can recover their wild character by genetic transformation if the introduced gene re-establishes the deficient function. However, sometimes the experimenter does not have a mutant corresponding to a well-known metabolic function. Such a mutant can be created by random introduction of small DNA fragments into the yeast genome of a wild type. When these fragments are inserted into a non-coding region (A), there is no consequent change. When they are inserted in the reading frame (B), there is a disruption of the gene, which is then no longer functional. The function may nevertheless be re-established by complementation thanks to the introduction of a functional gene.

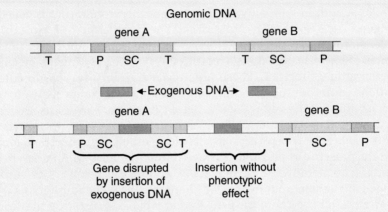

Gene disruption by random insertion of DNA

This technique of mutation by DNA insertion is also used in plants. The T-DNA of the Ti plasmid of *Agrobacterium* is a good candidate to knock out the gene in which this exogenous DNA is inserted at random. We can also target a precise point of insertion by placing in the T-DNA a sequence presenting enough homology with the gene sequence that one wishes to disrupt or destroy. This results in gene targeting. Finally, other researchers use transposons present in some plants to achieve the same goal.

There are many other relatively complex methods but they are rarely used.

When the gene is isolated and sequenced, it is possible to identify, i.e., understand, the exact role of the protein that it has coded for the synthesis of. There is already valuable information to be gained from knowing the deficiency that the mutant suffers before it finds its wild character, following the operation of a functional complementation. It is possible to go further and compare the sequences. There are, in fact, European and international data libraries (e.g., EMBL) in which any researcher who identifies a sequence, and who desires a recognition of his or her discovery, must deposit the sequence and specify, when possible, the function to which it is known to correspond. An accession number of the deposited sequence is attributed to the author. These libraries contain several millions of sequences. Biologists have at their disposal computer programs that can, in a relatively short time, compare the sequence they have identified with these millions of sequences in the library. They can thus find out the sequences that are either identical or very similar or even slightly different from their own, the organism from which this sequence has been isolated, the physiological function of the corresponding protein when that has been identified, and the published references to these sequences.

2.2. CONSTRUCTION OF CHIMERIC GENES AND SELECTABLE MARKER GENES

The DNA sequences corresponding to genes thus decoded can be used in their entirety. The promoters of these genes, however, are often subject, in their cells of origin, to controls that may not necessarily prevail in their acquired cells. The situation in the acquired cells is practically impossible to predict. That is why biologists prefer to use promoters that are relatively insensitive to their environment and capable of expressing strongly in the host cell. Thus, chimeric genes are constructed by placing the coding sequence between a strong promoter and a terminator known to be effective (Inset 2.8).

INSET 2.8

Transfer, expression and selection vectors

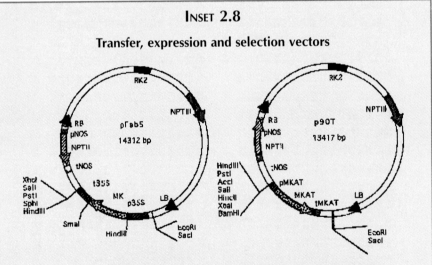

Two examples of vectors constructed on the same principle

At left, pFab5 designed to transfer the mevalonate kinase gene of baker's yeast into a target plant. At right, a comparable vector in which the coding sequence of the mevalonate kinase gene of *Arabidopsis thaliana* under its own promoter and its own terminator has been inserted.

The transfer vector is characterized by the presence of left and right borders of T-DNA between which the two genes have been placed in tandem: the expression gene and the selection gene. Each gene has the three essential parts: promoter, reading frame, and terminator.

In the example presented here, the expression vector comprises the strong promoter 35 S of the cauliflower mosaic virus (CaMV), followed by the coding sequence of the gene coding the synthesis of mevalonate kinase and then the terminator of the nopaline synthetase gene. This first gene is followed closely by the selection gene comprising the promoter of the gene coding for nopaline synthetase, the coding sequence of the aminoglycoside transferase II gene that confers kanamycin resistance on plant cells, and finally the terminator of the nopaline synthetase gene. The presence of borders favours the insertion of the entire exogenous sequence, called the "cassette", which from the regenerated plant will give rise to modified growth characteristics (the drawing shows a construction by Fabienne Lejeune, left, and Laurence Blanchard, right, in the vector Bin 19).

Example of coding sequence extracted from a cDNA library of *Arabidopsis thaliana* (constructed by F. Lacroute)

In the example of coding sequence extracted from the cDNA library of *Arabidopsis thaliana*, the sequence corresponds to the gene that codes for the synthesis of mevalonate kinase enzyme. This gene has been identified by functional complementation of a mutant yeast, then sequenced and deposited in the EMBL library under accession number 77793 by Catherine Riou. The enzyme that codes this gene catalyses the phosphorylation of mevalonic acid in the first stages of the isoprene biosynthesis pathway, which in plants leads to many molecules essential to their metabolism: e.g., cytokinin, gibberellins, abscisic acid, phytosterols. Note that more than the coding sequence whose beginning can be detected by the presence of an ATG, a non-negligible part of the promoter (P) is also present in the library. The terminator (T) is recognized by the presence of a polyadenylation region characteristic of messengers.

These genes are generally perfectly functional. However, their effect in the host organism is not always spectacular, and it is not always easy to distinguish, among the organisms that have been subjected to transgenesis manipulation, those cells with the incorporated transgene from those in which the transgene has not been integrated. To resolve this difficulty, we can simultaneously incorporate two genes placed in tandem: one being the chosen transgene and the other serving only to easily sort out the real GMOs from other organisms. This is the essential role of the selectable marker gene. Most often, it is a gene conferring resistance to a chemical substance such as a herbicide (round-up glyphosate, for example) or an antibiotic such as kanamycin or ampicillin. The known presence of a selectable marker gene indicates a very high probability of the presence of the transgene. One cannot theoretically or practically avoid the possibility that a rupture followed by a DNA recombination between the two genes will not survive, but this eventuality is rare enough that the process is not invalidated in principle.

The assembling of different parts of the gene necessitates the creation of sticky ends compatible with the ends of each DNA fragment so that these fragments assemble in the proper order (promoter, reading frame, terminator) and, for each of the genes, in the proper reading direction. The resulting construction is inserted in a plasmid that already has a bacterial selectable marker gene (ampicillin, for example) called the "cloning vector". These plasmids are incorporated into the bacteria that will multiply them. This is the phase strictly called the "cloning of the gene", which allows the biologist access to a sufficient quantity of transformant DNA, i.e., a few micrograms, to consider the following operation, gene transfer.

2.3. OPERATIONS OF GENE TRANSFER

Gene transfer is an important and delicate operation that comprises many modalities varying with the biological group to which the target of the transgene belongs. These techniques can be considered perfected today with respect to plants. However, they were first implemented in microorganisms, bacteria and yeast, before being adapted to more evolved life forms, animals and plants. An understanding of gene transfer in microorganisms is still indispensable since any transformation of the animal or plant cell requires the prior transformation of bacteria, and sometimes of yeast, particularly for selection of genes and construction of vectors. Let us quickly review the techniques used presently in microbiology and in culture of animal cells before addressing the focus of our study, plants.

2.3.1. In unicellular organisms

In unicellular organisms, bacteria and yeast, the techniques are today well established.

In bacteria, transformation is achieved by putting bacteria in a suspension in the presence of plasmids. Permeabilization of the membrane to incorporate the plasmid is ensured either by addition of calcium chloride or by thermal shock by electroporation.

In yeast, the initial technique is quite similar, but lithium salts are generally used. The DNA can be transferred by electroporation but the yeast cells must first be transformed into protoplasts by enzymatic lysis of the walls (using cutinase). The two techniques are used for transformation with nearly the same frequency.

2.3.2. In animals

The transformation techniques for animals are not fundamentally different from those used for genetic transformation of plant cells. Still, they are more restricted and the choice of technique depends on the objectives of the researcher and the cell type considered.

In animal transgenesis, the transformation can be done directly by microinjection and electroporation or indirectly through viruses. Unfortunately, there is no known bacterial vector, as is available for the higher plants. Moreover, animals do not have the facility of ontogenesis from numerous somatic cells, as do the higher plants. Only the female gametes and the very first blastomeres have some totipotentiality. This property disappears rapidly after the first few divisions of embryo cells.

Transformation by microinjection seems to be the most suitable technique for large cells such as ovocytes. The operation takes place under a microscope or an inverted microscope. The cell is held at the end of a small cannula by slight aspiration, while a microneedle, directed by micrometric vision, is introduced in the cell and if possible in the nucleus. A slight pressure on a piston propels the contents of the syringe, i.e., the DNA fragments containing the transgene accompanied by its selectable marker gene. This technique requires some training and its yield is never very high. About 100 microinjections a day constitute an excellent performance and not all of them are successful.

When smaller and more numerous cells are to be transformed, a particular electroporation method can be used that relies on slightly more complex equipment than that used in microbiology and especially on the use of complementary modules. The DNA can be incorporated without particular protection, but the biologist often relies on the fabrication of liposomes, a sort of artificial concentric membrane made up of phospholipids that contains the DNA fragments. The electric shock allows fusion of these artificial membranes with the plasma membrane of cells and the DNA is thus released in the cell (see Inset 2.13 below for the technique).

However, to reach the nucleus, the DNA must cross many barriers, including the endonucleases, enzymes with the sole aim of destroying them. The yield is not high, but some interesting results have been obtained.

Indirect methods of transfer in animals rely on viruses, which are capable of invading cells, carrying their information, and controlling the functioning of the genome of the host cells. However, the experimenter must take into account the specificity of the virus to its host cell as well as the relative harmlessness of the virus during its cycle of proliferation. The human type I adenovirus (AAV2; single-stranded DNA) seems to be a serious candidate for transgenesis because it is non-pathogenic and very poorly immunogenic. It must, nevertheless, be "tinkered with" to increase its specificity to the target. This seems to be possible by inserting in the capsid a protein that gives it the property of linking itself to certain cell types. It is also possible to use retroviruses, which have the advantage of carrying the transgene until its integration in the genome of the receiver cell. To transform ovocytes close to maturity, the spermatozoid can be used as a vector of DNA fragments.

2.3.3. In plants

The transformation techniques for plants are relatively varied and effective because the great majority of GMOs presently known are plants. Indirect transformation by the *Agrobacterium* system is by far the most widely used method but direct transformation techniques are also used because of their advantages. The two technologies are, in fact, complementary and each responds to specific objectives.

a) Indirect transfer methods: the Agrobacterium *system*
Agrobacteria are telluric, i.e., they live mainly in the aerated part of the soil. They are gram-negative aerobic eubacteria shaped like rods. In the spring, they may fix themselves on the crown of some Dicotyledons. The crown is the part of the plant corresponding to the junction of the root and the stem, a region usually located at the soil surface. Under the influence of the bacterium, it develops tumours and then necrosis. A section from the tumour shows an intensive multiplication of plant cells closely linked to a considerable proliferation of the bacterium. It is practically a cancerous tumour of the plant, but it does not spread. Generally, there is no metastasis in plants. The bacterial proliferation is maintained by the secretion of molecules that are unusual in a plant, the opines, which serve as food for the bacteria. This synthesis and secretion are induced by oncogenes situated in the transfer DNA, or T-DNA, which disturb and change the normal metabolism of the plant cell in favour of the synthesis of opines.

There are different species of *Agrobacterium*, of which the best known and most frequently used for transgenesis is *A. tumefaciens*, which harbours the Ti (tumour-inducing) plasmid, and *A. rhizogenes*, vector of the Ri (root-inducing) plasmid, inducing an intensive development of roots. In *A. tumefaciens*, there are two strains depending on the type of opine secreted: octopine and nopaline. The genes for synthesis of these substances are located in the T-DNA, while the genes to control their catabolism are found outside the T-DNA.

We have seen that it is the understanding of the molecular mechanisms of crown gall disease that inspired this mode of indirect transformation of somatic cells. These transformed cells are the source of a GMO. The T-DNA that has been disarmed of the Ti plasmid serves as vector. For it to serve as vector, it is necessary to conserve the two borders, left and right, of this T-DNA and insert in it the sequence or "cassette" containing the two genes: the useful gene and the selection gene under strong promoter. This cassette is generally constructed in a plasmid, which is replicated within a colony of *E. coli*. This plasmid is transferred into *Agrobacterium* by a conjugation between *E. coli* harbouring the plasmid and the bacterium. To facilitate the conjugation, it is sometimes necessary to use a third "helper" bacterium that favours the conjugation. This is called triparental conjugation.

In *Agrobacterium*, the plasmid can be transferred according to two protocols: cointegration and binary vector.

Cointegration vector corresponds to the union of disarmed Ti plasmid and an "intermediary" plasmid of *E. coli*, generally pBR 322 plasmid, in which the plant selection gene has been integrated. Most often, neomycin phosphotransferase gene type II or NPT II selection gene is used, which confers on the plant cell resistance specific to kanamycin antibiotic. This plasmid also possesses a cloning site on which we insert the desired gene to be transferred and a T-DNA fragment that serves to facilitate the recombination with the Ti plasmid of the agrobacterium. The vector plasmid is then multiplied a large number of times by *E. coli*. The vector is said to be *amplified*. It is then transferred by conjugation in an *Agrobacterium* with disarmed Ti plasmid. The homologies between the two regions of T-DNA, carried on the one hand by the transfer vector and on the other hand by the Ti plasmid, allow the integration of the intermediary plasmid between the two ends or borders of the Ti plasmid, i.e., in a position favourable to its ultimate transfer into the plant cell. Note that the recombined vector is finally larger than the original Ti plasmid, which is in fact a disadvantage for its successful transfer to the plant. Besides, in this technique, the origin of replication of the plasmid is found transferred and can thus easily replicate, a disadvantage that is not found in the binary vector technique. That is why the cointegration technique has practically been abandoned at present in favour of the binary vector technique, which is more recent and more effective (Inset 2.9).

INSET 2.9

Transgenesis by cointegration

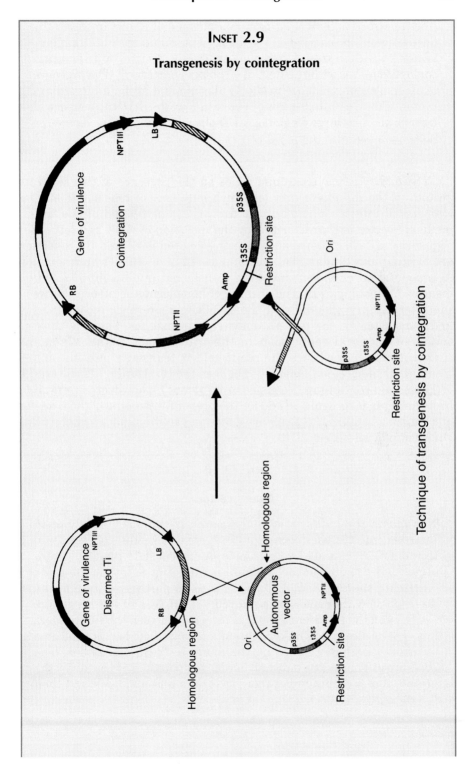

Technique of transgenesis by cointegration

The construction is done at first in a plasmid of *E. coli* in which the desired gene is inserted. This plasmid is called the intermediary vector. It will be transferred into an *Agrobacterium* by triparental conjugation, i.e., with the help of a "helper" bacterium. The presence of a homologous sequence in the Ti plasmid and in the intermediary vector allows an integration of the latter in the T-DNA, which will then form a very large recombinant Ti plasmid that could nevertheless be transferred into the plant cell.

The *binary vector* technique relies on the presence of two separate vectors in the *Agrobacterium*, of which only one, of reasonable size, is transferred into the plant cell. At first, the construction of the "intermediary" vector is characterized by the insertion of right and left order sequences of T-DNA between which the desired gene and the selection gene are found in tandem. This vector plasmid is amplified by introduction in a population of *E. coli* and then transferred into an *Agrobacterium* disarmed by conjugation. The absence of homologous sequences between the intermediary plasmid and the Ti plasmid prevents recombination and, consequently, the two plasmids remain distinct. There is thus no source of bacterial replication in the transformed plant. When the *Agrobacterium* parasitizes the plant cell, it transmits only the intermediary plasmid region contained between the two borders. The presence of Ti plasmid containing the genes of virulence is, however, essential to bacterium-plant recognition and for the proper functioning of the transfer mechanisms because the virulence genes can act from far away on the T-DNA boundaries (Inset 2.10).

INSET 2.10

Binary vector

Nowadays, the binary vector technique is the most widely used indirect gene transfer technique in plants. Like the cointegration technique, it requires two plasmids, a disarmed Ti plasmid and a vector called "autonomous" because it cannot at any time integrate itself with the Ti plasmid. To be able to be partly transferred, the autonomous vector must be equipped with right and left borders of T-DNA and the transfer cassette is inserted between the two borders in a cloning site or polylinker. This site is constructed so as to be able to insert the DNA cleaved by the many restriction enzymes. The presence of Ti from one side and border to another is essential for the processes of recognition between bacterium and plant cell and then the transfer can occur.

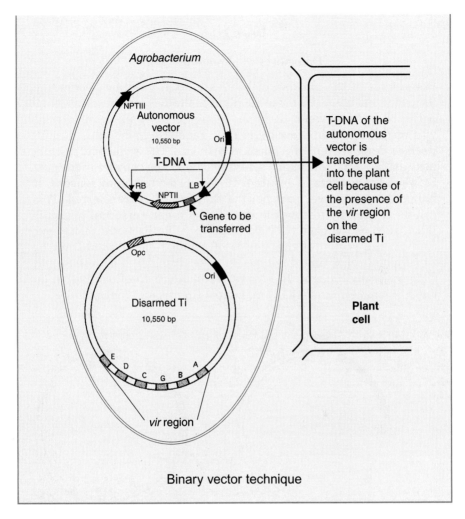

Agrobacterium

NPTIII
Autonomous vector
10,550 bp
Ori
T-DNA
RB LB
NPTII
Gene to be transferred
Opc
Ori
Disarmed Ti
10,550 bp
E
D
C G B
A
vir region

T-DNA of the autonomous vector is transferred into the plant cell because of the presence of the *vir* region on the disarmed Ti

Plant cell

Binary vector technique

The binary vectors available to biotechnology researchers are relatively numerous. Some vectors are more commonly used than others, such as vector Bin 19 or Bin plus.

These vectors possess:

- a source of replication,
- the right and left borders of T-DNA between which the kanamycin resistance gene is placed under the strong promoter of CaMV 35 S and the nopaline synthase terminator, and
- a multiple cloning site that allows introduction of the desired gene into the vector for transfer.

A second selection gene with a bacterial antibiotic (ampicillin, type II aminoglycoside transferase, erythromycin) external to the T-DNA allows selection of bacteria carrying the plasmid (Inset 2.11).

INSET 2.11

Derivatives of plasmid pBR322

Derivatives of the plasmid pBR322 are often used for selection in bacteria. For gene transfer in plants, the best known is Bin 19 perfected by M. Bevan in 1984 but sequenced only ten years later. It was modified and improved and thus gave rise to Bin-plus. Bin 19 itself is derived from the plasmid pBI 121. These plasmid vectors possess "bacterial" or "plant" selection genes. The transfer of the part between the borders is ensured. However, for unknown reasons, it happens that a larger part beyond the borders is transferred. This can be verified by testing the presence of a gene external to the T-DNA, such as the gene NPT III responsible for bacterial resistance, or even the region of the source of replication. It would not be surprising to find that this type of "wide" transfer is more frequent than generally thought. In France, for instance, the Commission du Génie Biomoléculaire (CGB) mandates this type of control.

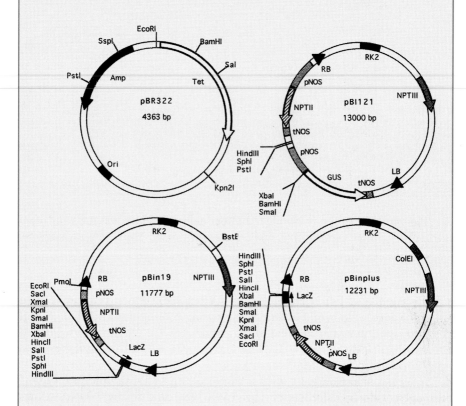

Examples of plasmid vectors frequently used in transgenesis

For the desired gene to be clearly expressed, the coding sequence of a strong promoter is placed upstream. The promoter expresses itself beyond the control of the host cell and consequently has constitutive expression in all the tissues and organs of the plant, no matter what its stage of development. This promoter is taken from the cauliflower mosaic virus (CaMV), which codes for the synthesis of a capsid protein. This is the CaMV 35 S promoter, capable of implementing the high-level transcription of the gene placed downstream if this last sequence is placed "in phase" with the promoter and no other stop codon will truncate the transcription.

INSET 2.12

Antisense vector

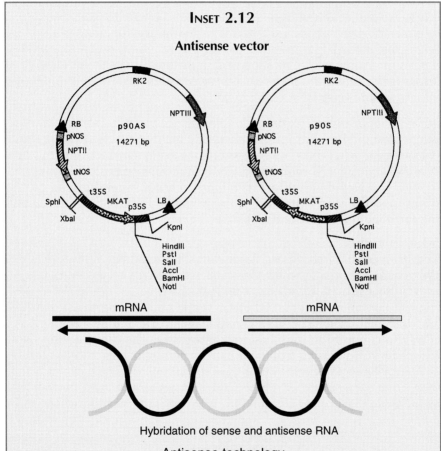

Hybridation of sense and antisense RNA

Antisense technology

The experimenter sometimes wants to reduce or block the functioning of a gene instead of overexpressing it, without, however, destroying or artificially mutating the gene. This can be done by means of antisense technology. The antisense vector is constructed so as to insert, downstream of a strong promoter, the reading frame of the gene oriented in the reverse direction. This gene codes the synthesis of messengers that are thus complementary to messages transcribed by the gene in the normal direction. It is thought that the two categories of messenger can thus hybridize, which means that they can be used by translation enzymes. No protein is synthesized, and the gene is said to be "extinct".

On the other hand, sometimes it is necessary to reduce or even annul the level of transcription of a gene. This technology of transgenesis is called "antisense strategy". The coding part or even simply one of the short fragments is placed in inverse position under the control of the strong promoter, which leads to the formation of messengers complementary to those transcribed by the correctly positioned gene. These two mRNA can couple to form double strands and they cannot be used by the translation enzymes. In principle, no protein can be synthesized by these genes if the two types of RNA are produced in equivalent quantities. A low level of expression may result in an incomplete coupling of two classes of messengers, especially if the antisense category is produced in smaller quantity than sense RNA (Inset 2.12).

b) Direct methods of transfer

Even though they are less often used, techniques of direct genetic transformation are quite widespread, particularly for the transformation of Monocotyledons (Graminaceae) or Dicotyledons that are less sensitive to *Agrobacterium*. They are common to techniques applied to the transformation of animal cells: microinjection of DNA, somatic fusion and the liposome technique, and biolistics.

MICROINJECTION

The *protoplast* is the most favourable material for microinjection. Protoplasts are cells in which the walls are removed by maceration in an enzymatic solution. To achieve this result, the cells are separated from one another following digestion of the pectic wall by a pectinase solution. They are then, and sometimes simultaneously, treated by a combination of cellulases until the wall dissolves completely. The osmotic pressure of the medium is adjusted by addition of various salts or mannitol (around 13%) in order to prevent the bursting of the protoplast. The major difficulty then encountered is proper repair of the nucleus because it is often masked by the chloroplasts and pushed laterally by the large central vacuole (Inset 2.13).

INSET 2.13

Microinjection of DNA in plant protoplasts

The microinjection is done using a micromanipulator mounted on an inverted light microscope. In the first place, we look at the pneumatic control of the micromanipulator, which allows a very fine approach of two microneedles, one to hold the cell to be transformed by means of a slight aspiration exerted by the syringe located at left, the other

---

to inject the DNA using a slight pressure exerted by the syringe at right. After injection, the protoplast is liberated and cultured in an appropriate medium.

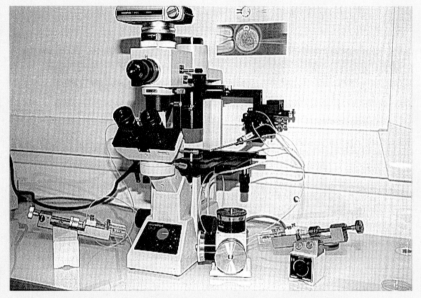

Apparatus for microinjection of DNA in plant protoplasts

SOMATIC FUSION AND THE LIPOSOME TECHNIQUE

The technique used to transform by somatic fusion is directly inspired by that used for animal cells using liposomes, described in section 2.3.2 above. In plants, it can be applied only on protoplasts. The fusion of protoplasts with each other or with liposomes containing a transformant DNA can be facilitated by certain substances such as polyethylene-glycol at high concentration (up to 30%). It is also possible by electric shock from an electroporator adapted to eukaryote cells. The electric shock is sometimes preceded by an alignment of protoplasts following the application of a high-frequency current (Inset 2.14).

INSET 2.14

Electrofusion

The same inverted microscope used for microinjection can be equipped for electrofusion (electroporation) of plant protoplasts. In the first step, below the plate, a generator of high-frequency electric current is used to align and bring the protoplasts together. Above, a generator of continuous current supplies to an electroporater an electric shock of several hundreds of volts for a few milliseconds (or microseconds, as desired) via electrodes placed in a small petri dish containing the protoplasts. This electric shock makes the plasma membrane permeable and momentarily allows the entry of DNA molecules suspended in the culture medium.

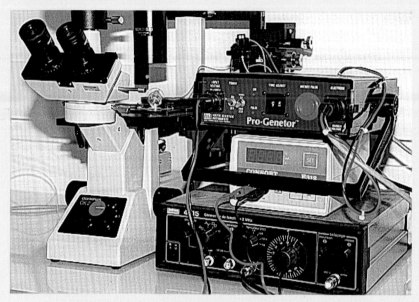

Apparatus for electrofusion preceded by the application of high-frequency current

Genetic transformations have been realized through this technique notably in representatives of the Solanaceae: tomato-eggplant fusion by G. Ducreux and colleagues at the University of Paris-XI. Results corresponding to true genetic transformations have been obtained by protoplast fusion with liposomes containing DNA fragments (Inset 2.15).

Inset 2.15

Liposome technique

The lipofection or lipotransfection technique consists of encapsulating DNA fragments carrying the genes to be transferred in spherical structures called "liposomes" made of lipid or phospholipid double layers resembling plasma membranes. These liposomes fuse, spontaneously or under the influence of PEG in the medium, with the plasma membranes of protoplasts prepared from a culture of isolated cells and liberate their contents in the cytoplasm of the plant cell. A certain number of these DNA molecules can be destroyed by nucleases. A minority of them reach the nucleus and can integrate themselves into the cell genome to be the source of a GMO.

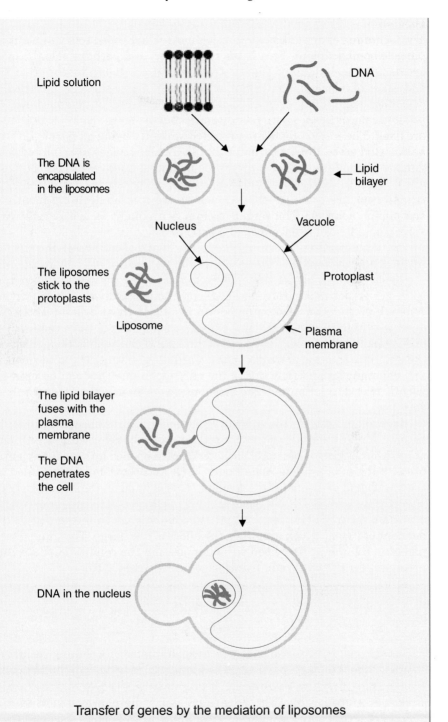

Lipid solution

DNA

The DNA is encapsulated in the liposomes

Lipid bilayer

Nucleus

Vacuole

The liposomes stick to the protoplasts

Protoplast

Liposome

Plasma membrane

The lipid bilayer fuses with the plasma membrane

The DNA penetrates the cell

DNA in the nucleus

Transfer of genes by the mediation of liposomes

BIOLISTICS

The technique of direct transformation most widely used today in plants is particle bombardment. This is the most recent and perhaps also the most spectacular technique of genetic transformation. It originated from the encounter of a biologist and a ballistics expert; hence the term *biolistics* to designate this highly original technique. It consists of bombardment of isolated cells or tissues with metallic microbullets on which DNA molecules are fixed. The microbullets are projected violently using a type of gun that has existed since 1987: a carabine 22 LR. The cartridges are charged with gunpowder and projectiles represented by microbullets of tungsten or gold about 1 μm in diameter. The DNA fragments containing the genes to be transferred are stuck to the surface of the microbullets by means of spermidine solution. The target is often a fragment of leaf parenchyma spread on a disc of filter paper in a petri dish.

In the beginning, "shooting" led to the death of cells receiving the full blast of the projectiles. However, at the periphery of the impact, the superficial cells, such as the epidermal cells, may be merely injured by some projectiles and survive at least long enough to divide. The surviving cells are placed in a selection medium to isolate those cells in which the gene was able to penetrate and express itself, even if the expression is only *transitory* because of the non-integration of the gene in the genome of the cell. Note that the projectiles can eventually penetrate the mitochondria and chloroplasts and that this is the only technique that can be used for genetic transformation of these genomes, even though the success rate is very low.

Nowadays, the gun has been replaced by a pressurized cylinder of helium. A regulator allows this propellant gas to be released at a pressure of 7 to 8 bar, which by means of an electric hatch opened by the experimenter projects the microbullets coated with DNA and provisionally retained on the metal grill of a small filter opening. This is fixed on the upper part of a watertight enclosure with a glass opening in which a vacuum pump creates a partial vacuum. The residual pressure is only 25 to 30 mmHg at the moment of firing, which amplifies the effect of discharge. The gun is itself placed in a laminar flow hood so as to maintain the conditions of sterility necessary for the culture of transformed cells (Inset 2.16).

INSERT 2.16

Biolistics

Biolistics is a technique of genetic transformation combining certain properties of biology, particularly genetic engineering, with ballistics, because it first used a firearm. It originated in 1987 and has since evolved considerably. The apparatus shown here corresponds to a state of technology dating from 1997 and still frequently used today. It involves a gun with microbullets of tungsten.

Apparatus for genetic transformation by the biolistic technique

Petri dish

The enclosure of the device has thick walls, the front wall made of glass, and a nozzle with a filter opening in the upper part. This is linked to an external airlock by an electric valve that can put the filter opening in communication with the lock. The lock itself is connected to a pressurized cylinder of helium with a regulator set at

7 or 8 bar. The pressure in the container can be considerably reduced by a vacuum pump (around 25 mmHg). This pressure is controlled by a nanometer visible on the side. The microbullets of tungsten are coated in a layer of spermidine, which allows them to fix reversibly the DNA molecules to be transferred and then placed on the filter opening grill. The target is made up of tissue fragments (leaf parenchyma, for example) arranged on a filter paper in a petri dish placed on a platform located in the container at a variable distance from the filter opening.

Pressure is applied on an interrupter to open the valve releasing the compressed gas, which projects the microbullets on the target. The tungsten particles penetrate more or less deeply into the cells, lodging at random in the cytoplasm, the nucleus, the plastids, or the mitochondria. The transgenes can then express themselves transitorily or, much more rarely, integrate themselves in one of the genomes of the cell (nuclear genome or genome of semi-autonomous organelles).

The results obtained with this technique may be of three kinds:

1. The DNA carrying the gene does not integrate in the genome of the host cell and expression of the gene is only transitory. This situation is useful, despite the non-integration, because it allows us to study the functioning of the gene, its translation products, and its control. The cell is nevertheless not truly considered genetically transformed since it cannot transmit its acquired property to its descendants.

2. The DNA carrying the transgene penetrates the nucleus and, finding a region showing high homology of sequences, integrates itself in the genome of the host cell by a double cross-over. The integrated character thus becomes hereditary and the cell is genetically transformed. This technique has been widely used for transformation of maize. Generally, it seems to be an interesting alternative for all the plants insensitive to *Agrobacterium*.

3. The DNA is introduced in the chloroplasts or mitochondria but not in the nucleus, and the genetic transformation applies only to these organelles. This case is quite rare but has yielded results especially for the introduction of genes for antibiotic resistance, a property that is manifested when the transformant DNA has been introduced in the genome of cell organelles.

AGROLISTICS

Agrolistics is a combination of two transformation techniques that combine biolistics with indirect transformation using agrobacteria (Inset 2.17). Indeed, biolistics continues to evolve and some researchers have sought to combine several techniques to obtain the genetic transformation of some species that are less accessible to transformation by any one of the techniques cited. In agrolistics, the gun was used to simply induce micro-

INSET 2.17

A manipulation of genetic transformation in tobacco

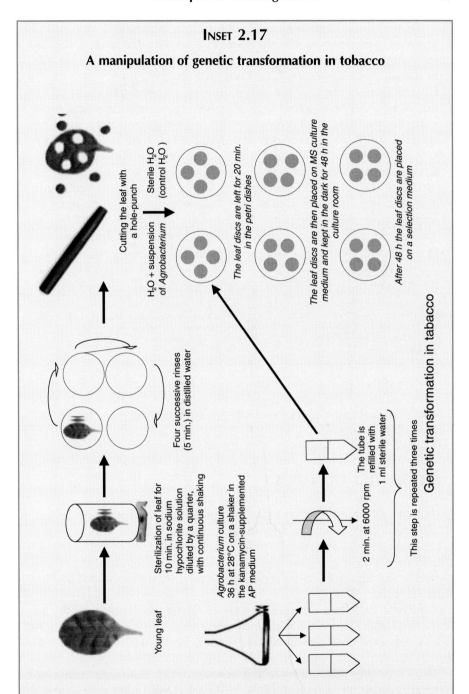

Genetic transformation is done entirely in antiseptic conditions under a horizontal laminar flow hood. On the day before the manipulation, the *Agrobacterium* culture is prepared from a colony preserved at −80°C. These bacteria carry a Ti plasmid that has between the

T-DNA borders the gene to be transferred, in tandem with a gene for kanamycin resistance. The *Agrobacteria* are cultured at 28°C for 24 to 36 h, in a specific culture medium (AP medium) containing the antibiotic, here kanamycin, and put on a shaker.

On the morning of the procedure, young leaves of the plant to be transformed are sterilized by a solution of sodium or calcium hypochlorite (commercial concentrated bleach diluted with distilled water, 1 vol./3 vol., is perfectly suitable) for 1 to 10 min., followed by a quick rinse in 70% alcohol and four successive rinses in sterile distilled water. Meanwhile, a suspension of *Agrobacterium* is prepared in sterile water after centrifugation. The leaves are cut into small discs using a hole-punch. The leaf discs are placed for 20 min. in contact with the suspension of *Agrobacterium* in the sterile petri dishes and then cultured on a nutrient solution of Murashige and Skoog (MS) medium, for 48 h at 24°C and in the dark. During this period, the bacteria invade the leaf discs through peripheral cells damaged by the cut. At the end of this period, the discs are transferred to a culture medium containing the antiobiotic that will select the transformed cells by destroying all the non-transformed cells. In parallel, an operation is done without *Agrobacterium* that serves as control.

injuries in cells that were subsequently exposed to *Agrobacterium* carrying the desired gene. The micro-injuries favoured the invasion of tissue by *Agrobacterium* and the yield of the transformation was found to be considerably increased.

2.4. ESSENTIAL CONTROLS

Even if the parameters of transfection operations are totally respected, a stable transgenic organism may not always be obtained. The researcher must study several controls to verify the effective transfer of the gene, its transcription into a messenger that can be used by the appropriate enzymes, and finally the synthesis of proteins that are functional and have reached maturity.

The first precaution thus consists of verifying the presence of the gene in the host cells. There are various techniques for verification but the two most commonly used are Southern blot and polymerase chain reaction (PCR).

Southern blot (named after the scientist E.M. Southern) consists of cleaving the double-strand DNA by restriction enzymes. The DNA fragments are then separated by electrophoretic migration in an agarose gel, denatured, and transferred on to a nitrocellulose membrane. This nitrocellulose membrane is then put in the presence of a specific complementary probe of the desired sequence, with which it will hybridize if they are complementary. The probe is detected either by its being labelled with a radioactive element or by its luminescence. One or the other of these properties is expressed, even if present in traces, and is clearly visible on a photographic film (autoradiography).

If no specific probe is available but the nucleotide sequence of the gene is known, it is possible to synthesize two small sequences of specific DNA called "primers" because they can amplify up to a thousand copies of the gene sequence, if the gene is present, by means of a PCR reaction (see Inset 2.18). This same PCR cannot result in any amplification of a gene in a cell that does not possess the gene sequence in question. To go further with the control, we must cleave the gene sequence at a well-defined locus, which will determine the presence of two fragments of known molecular masses that can be separated on a gel. The same cleavage can be made on amplification products in the host cells and we can verify that we have the same two fragments that were expected in theory (Inset 2.18).

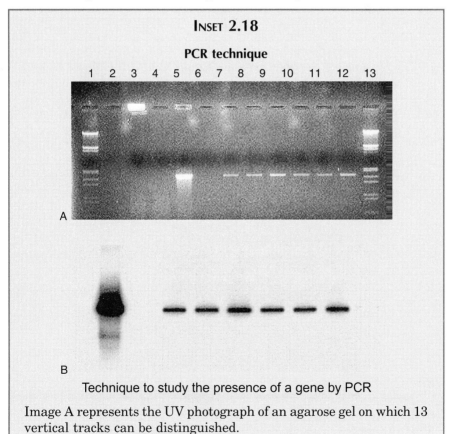

INSET 2.18

PCR technique

Technique to study the presence of a gene by PCR

Image A represents the UV photograph of an agarose gel on which 13 vertical tracks can be distinguished.

Tracks 1 and 13 represent controls of molecular masses on which combinations of DNA of known molecular masses have migrated. The following observations are made:

- Track 2 is an H_2O control.

- Tracks 3 and 4 are controls with a single primer (track 3, first primer; track 4, second primer).

- Tracks 5 to 12 represent DNA extracted from six different plants: the plants corresponding to track 6 do not have the transgene and thus have not been transformed. The others are all genetically transformed.

Image B corresponds to the transfer of DNA from the gel to the nitro-cellulose membrane. These denatured DNA have been hybridized with a specific probe of a desired gene (radio-labelled complementary DNA). The membrane has been left in contact with a photographic plate that has registered the radioactivity at the place where the probe was able to hybridize with the desired DNA. This autoradiography confirms the presence of the gene. It is observed that the amount of the transgene (number of copies present in the plant) can be "quanti-fied" on the basis of the intensity of labelling registered on the auto-radiography.

The presence of the gene does not signify that it functions. It is often necessary to find the presence of messenger RNA corresponding to the transcription of the gene. In practice, the technique used is very similar to that of Southern but the messengers are identified by hybridization with a cDNA probe corresponding to the coding sequence of the gene concerned.

The technical resemblance to Southern blot has resulted in the name *Northern blot* for this technique (Inset 2.19).

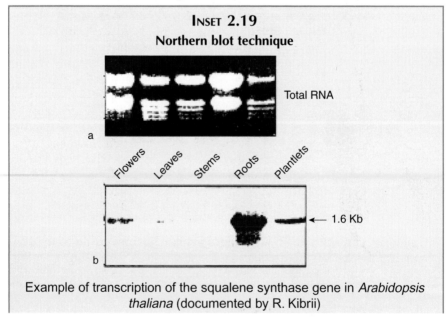

INSET 2.19
Northern blot technique

Example of transcription of the squalene synthase gene in *Arabidopsis thaliana* (documented by R. Kibrii)

> a. Total RNA on the agarose gel after staining with ethidium bromide (EtBr) and illumination with ultraviolet.
> b. Autoradiography after transfer on a nitrocellulose membrane and hybridization with a specific probe of the radiolabelled gene.
>
> It can be observed that this gene is expressed very strongly in the roots, less strongly in the flowers, and not at all in the leaves and stems. It is also expressed in the young plantlets germinated from seeds.

Similarly, the presence of specific messengers does not necessarily imply the presence of functional proteins. It is advisable to seek and identify the proteins either by using specific antibodies or, if enzymatic proteins are involved, by verifying their catalytic efficiency. The reaction with the antibodies occurs after migration of the proteins on an SDS-PAGE gel followed by a transfer on to nitrocellulose membrane. This technique, by analogy with the two preceding ones, is called *Western blot.*

2.5. STABILITY, HERITABILITY AND DISPERSAL OF THE TRANSGENE

To be certain of having created a true GMO, it is necessary to establish controls at three levels of integration of the transgene: stability, heritability, and dispersal. It is true that, in practice, all these controls are not always implemented and the experimenter often stops with phenotypic criteria considered to be necessarily linked to the presence of the transgene. It is also necessary to ensure the stability of the transgene over the course of cellular events marking sexual reproduction: meiosis and fertilization. For this, the tests indicated above are in fact often done on young seedlings germinated from seeds produced by plants derived from the regeneration of transformed cells. Sexual reproduction also results in the dispersal of the transgene, especially with species that practise allogamy or interspecific hybridization. We will see that this is the case with colza or rape, for example, which can hybridize in nature with cabbage, wild radish, and mustard. There are thus significant risks of dispersal that may induce regulatory commissions to prohibit cultivation trials of these GMOs.

These controls can be further complemented by research on toxicity due to the presence of the transgene.

Chapter **3**

Research Programmes and Results

3.1. OBJECTIVES OF RESEARCH PROGRAMMES

3.1.1. Overall objectives

The reasons that impel researchers to use genetic engineering techniques to create GMOs are extraordinarily varied. Their activities have accordingly ranged from preoccupation with the most noble aims of purely scientific knowledge to commercial research for large multinational corporations. One particularly revealing development over the past 25 years has been common to all these aims: the reorientation of research projects of many biology laboratories. Genetic engineering very quickly was found to be a powerful and effective tool that could provide answers in some fields of research, of widely varying orientations, that have been stagnant or languished in dead ends. Research structures have been set up—laboratories, institutes, and centres—combining teams that not only have common research objectives, but also use the same molecular biology techniques, thus improving their access to costly materials such as oligonucleotide synthesizers or automated DNA sequencers. It is therefore not easy to classify simply and objectively the true reasons that have motivated operations of transgenesis in all the taxonomic groups representing the living world. Still, let us try to highlight, essentially for plants and their precious and indispensable auxiliaries, the microorganisms, the major objectives that are more or less acknowledged to be at the source of GMOs, objectives that have given rise to many of our concerns.

3.1.2. In microorganisms

Microorganisms have been the object of a large number of genetic engineering experiments for two essential reasons: either for themselves or as valuable and indispensable auxiliaries in the genetic transformation of higher animals and plants. For plants, bacteria such as *E. coli* and *Agrobacterium tumefaciens* are most often used, while yeast (*Saccharomyces cerevisiae*) represents an outstanding model. It is essential to explain here more precisely the relationships of these microorganisms with plant transgenesis.

a) Bacteria

Microorganisms, notably *Escherichia coli*, were the first to be subjected to research on genetic transformation, which was essentially dedicated to pure research and teaching. Such pure research ended in highly effective control over functioning of the prokaryote cell, its genome, and its plasmids, but it also found applications in medicine, veterinary medicine, pharmaceuticals, parapharmaceuticals, and agro-food. No GMO is presently authorized to be produced commercially for direct use in food for humans but useful molecules produced by GMOs are widely used in the food industry and in pharmaceuticals. Such molecules are enzymes, vitamins, or sweeteners produced by the fermentation of microorganisms that are mutants or have a modified genome. These molecules are subsequently involved in the manufacture of food, pharmafood, and vaccines and therapeutic proteins. The media containing the molecules must always be controlled and kept totally free from bacteria or their DNA. The enzymes thus produced can legally be used in the manufacture of beverages (beer, wine, alcohol), bread, cheese, or glucose or maltose syrup, especially in processes of fermentation or release of intermediary metabolites.

There is a difficulty in legislative orders in separating naturally mutant bacteria, i.e., bacteria in which the genome is modified spontaneously and inexplicably, from bacteria that became mutant for the same function subsequent to a directed mutation especially by genetic disruption. Even though the two types of bacteria may be phenotypically identical in every point, the first may be authorized and the second prohibited! This is true, for example, of certain lactic bacteria involved in the manufacture of yoghurt.

Let us also note that the use of GM bacteria has been authorized in a kit for detection of traces of antibiotics or products of degradation of these antibiotics in milk. The same is true for the manufacture of genetically engineered insulin designed to treat diabetics or even the human growth hormone. In the last case, the intervention of genetically reprogrammed bacteria seems infinitely less dangerous than extraction from the pituitary glands of cadavers, a technique that was earlier used and often had seriously disappointing results.

In animal feed, the situation is presently the same as for human food, because of the food chain. Authorized enzymes have been associated mostly with improved food digestibility (e.g., cellulases, xylanases, glucanases).

Authorization is much easier to obtain for industrial applications. Restrictions pertain essentially to the prohibition of dissemination of GM bacteria that have been involved in fermentation and various industrial procedures generally to a natural environment.

b) Yeast and other mycophytes

Yeast as well as a certain number of microscopic filamentous fungi are microorganisms but belong in fact to the Eukaryotes. Their cell function is thus very different from that of bacteria and is similar to that of animals and plants (particularly animals, because they do not photosynthesize). Among these organisms, baker's yeast, *Saccharomyces cerevisiae*, and some other kinds of yeast (such as *Schizosaccharomyces*) are particularly involved in transgenesis. Certain yeast species that belong taxonomically to the Basidiomycetes, such as *Phaffia rhodozyma*, have become model organisms for the controlled production of alimentary carotenoids. Yeast is a very easy material to use, since its genome was entirely sequenced in 1996 and its metabolic characteristics are relatively well known. There is also a remarkable collection of mutants whose metabolism is blocked at various levels, so that there is a wide choice of material for research and identification of a large number of genes that can subsequently serve as probes for plant gene research. Considering the importance of yeast in food fermentation processes, we can easily see why it became subject to genome modification to function in precise conditions and specifically targeted activities.

To meet the demand for a large number of mutants, artificial yeast mutants were created by random DNA insertion.

3.1.3. In plants

There are many reasons that motivated the orientation of some research laboratories toward the creation of transgenic plants. The first reason, historically as well as in terms of scientific interest, was essentially the desire for a thorough understanding of the plant genome and its modes of expression. This subject required a high-performance technical approach for the comprehension of metabolic pathways, enzymatic reactions, and messengers of all kinds that determine the plant's growth, development, and reproduction. The process normally begins with a description of the structures involved in the phenomena. Then there is a search for physiological mechanisms and correlations, and finally an understanding of the primary causes, i.e., in biology, the molecular mechanisms controlled by the genetic information in the cell. That is why many researchers have invested time in molecular biology and genetic engineering and more specially in the transgenic approach, which is the most effective form of these technologies. Such research could be confined to the laboratory, except when large-scale, outdoor trials are required. This process in its fundamental aspect must never be limited by moratoria of some kind that would inflict a highly detrimental delay on the progress of scientific research.

These fundamental aspects of research can be easily understood by the public, at least when the genetic transformations involve only unknown plants or those classified as weeds, such as *Arabidopsis thaliana* (Inset 3.1).

INSET 3.1

A small Brassicaceae

Arabidopsis thaliana

The small Brassicaceae *Arabidopsis thaliana* has become, in less than 20 years, one of the most important plants for researchers after having had only a modest mention in the weed treatises of agronomists. Today it is the model plant for genome research particularly because of the very small size of its genome, its restricted number of chromosomes ($2n = 10$), the facility with which it can be cultivated in laboratories, its 5 to 6 possible generations in a year, its many known mutants, and its exceptional proliferation rate. Its genome is entirely sequenced and the complete sequence was published in December 2000.

Transgenesis in tobacco is also quite well tolerated by the public, even though the plant is a commercial crop, because it is not actually eaten. Tobacco, because of its remarkable regeneration capacity, is often used in transgenesis experiments. It is thus a peculiar case.

In contrast, the application of transgenesis in plants that may end up in our food is far from being acceptable to people. It is understandable, however, that there were great temptations to be able to test research findings on plants interesting to humans in terms of food or economy. Two sectors are traditionally considered essential in the agronomic domain for experimentation: productivity, often defined as the yield per hectare, and plant protection against diseases and pests. Many other sectors of agronomy, horticulture, and silviculture were also the object of research in plant transgenesis. Let us look at some examples of programmes and results in GM plants.

3.2. ACHIEVEMENTS IN FUNDAMENTAL RESEARCH

Some species can be considered "model plants" because they are the basis of experiments in genetic transformation but they themselves are not an ultimate object. Tobacco (*Nicotiana tabacum*) was used for its remarkable regeneration capacity and wall cress; *Arabidopsis thaliana* because it had the smallest genome of the plant Eukaryotes. These are the two examples that we have already cited because they represent the two best-known cases. Among trees, the poplar (*Populus nigra* or *alba*) has similar advantages. It is also characterized by a small genome and is amenable to *in vitro* culture. For trees, the term *genetically modified tree* or GMT is used. With respect to cereal plants, rice is considered a model plant by certain researchers or for certain aspects of research, especially because of the size of its genome, which is one sixth the size of the wheat genome. At the same time, it is the object of applied research oriented towards its improvement according to purely agronomic criteria. It would be more exact to consider it a peculiar case. To date, the other groups, such as mosses (except one), ferns, and conifers in the wider sense, do not include species that are considered model plants.

Fundamental research in transgenesis aims for the most part to better understand the functioning of the cell through its major metabolic pathways and its exchanges with the external medium. The pathways of synthesis of fatty acids, isoprenoids, and especially carotenoid-type pigments and hormones of various origins are most often studied. We can also cite the processes that result in the excretion of secondary metabolic products or that are involved in cell division, particularly the role of cyclins. Let us cite, among the most recent studies in this field, the research on cyclin D, which controls the G1 phase of the cell cycle and reduces plastochron. It is possible, by controlling the activity of the gene, to control the morphogenesis and organogenesis of transformed plants.

At the organism level, transgenesis could involve the entire plant or only some organs. For the whole plant, the choice pertains to the use of a strong promoter, such as CaMV 35 S, to trigger the expression of the gene under any conditions. When only some organs are involved, a specific promoter of a gene specially functional in the organ or a promoter that is inducible by a particular signal would naturally be preferred. It is this last process that needs to be developed: a gene specific to roots, leaves, flowering, or formation of fruit or seed. These organs constitute anatomical or functional entities whose development or functioning could be privileged. From the perspective of metabolite transport, the organs that are taken into consideration are generally evaluated in their source/sink aspect.

In fundamental research, the leaf is often used because it is easy to remove and because the removal of a few leaves has a minor effect on the overall growth of the plant. The palisade parenchyma of the mesophyll is, moreover, the tissue that often regenerates the best. This property is probably linked to its essential role in carbon metabolism and due to the intensive migration of metabolites in this tissue.

The ontogenesis of the flower and of its verticils of organs, the control of petal colour, and, more recently, the sex determination of reproductive organs, especially in dioecious plants, are today a major focus of interest. For these studies, the model plants are wall cress, snapdragon (*Antirrhinum majus*), and petunia. In these plants the functioning of the chalcone synthase gene has been blocked by antisense technology and petunias with only white flowers have been obtained. The transfer of the gene for dihydroflavone-4-reductase of maize into petunia made it possible to obtain GM petunias with red flowers. With respect to stamens, the researcher is essentially interested in the sporogenic tissue, its role in the genetic expression of fertility or sterility through reactions of pollen compatibility, or even its maturation in relation with the nourishing role of the *tapetum cells*. The pistil and particularly its stigmatal part have also been the object of experimentation. These results are of obvious interest in horticulture.

An understanding of molecular mechanisms that control fruit maturation is useful in fundamental terms as well as in applications, especially for the storage, conservation, and transport of fruits. The control can be exerted through regulated production of ethylene, which induces and synchronizes their maturation, or, following an antisense insertion, by blocking the activity of pectinases that normally hydrolyse the intercellular pectic cement between the cells and cause softening of fruit at maturity. Control of fruit development and maturation by transgenesis has many applications in the agro-food and health fields (e.g., vaccination).

3.3. ACHIEVEMENTS IN APPLIED RESEARCH

In France and western Europe, the use of GMOs in agriculture is still extremely modest and often limited to "field trials". In the United States and more recently in some South American countries (Argentina, Brazil), GMOs are widely cultivated and represent 25 to 50% of production for soybean, maize, and cotton. At the end of the 20th century, the areas planted with GMOs in these countries exceeded 40 million ha. Genetically modified crops have become popular in developing countries. Let us look at some sectors in which GMOs can be used.

3.3.1. Agronomy

a) Selection

Techniques for selection and improvement of cultivated plants rely on hybrid crosses followed by individual selection from the descendant populations. These techniques are a direct application of Mendelian laws that were perfected in practical terms by R. De Vilmorin in the beginning of the 20th century. They are still valued for their performance but have always been time-consuming and costly. Various improvements designed to shorten the delays (use of male sterility instead of manual emasculation, the completion of several crop cycles a year by alternation between the northern and southern hemispheres) have been widespread and today constitute what we call conventional selection techniques. Diploidization of haploids is an additional technique that has been appreciated especially in cereals, colza and other Brassicaceae (Inset 3.2). It consists of obtaining haploid plants by culture of anthers or microspores and then doubling the chromosomal number from regenerating cells using colchicine, to obtain 100% homozygous fertile plants.

INSET 3.2

Creating a haplodiploid

To obtain a haplodiploid, we start with spores, generally microspores (androgenesis), before they develop into gametophytes (pollen grains). The microspores are extracted from the closed floral bud and cultured on an appropriate nutrient medium. Contrary to their normal development into the pollen grain, which occurs by means of a highly asymmetrical mitosis, the division in this case is perfectly symmetrical and yields 2 cells, then 4, then 8, and so on, as does the zygotic embryo. The embryoid thus formed develops into a complete haploid plant and is normally sterile because it cannot undergo meiosis to form spores. To overcome this disadvantage, the plantlet must be treated at a very young stage of development with colchicine, which at precise doses makes it possible to double the chromosome number. Very frequently, new plants must be regenerated from tissues in which the cells have become diploid, which often does not represent the majority of cells. The plants thus obtained are obviously 100% homozygous. Even though a delicate manipulation is involved, many

---

varieties of cultivated plants, particularly in Gramineae, have been obtained in this manner.

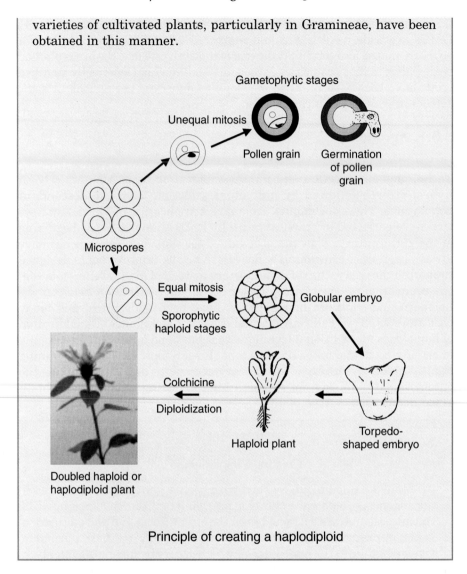

Gametophytic stages

Unequal mitosis

Pollen grain   Germination of pollen grain

Microspores

Equal mitosis

Sporophytic haploid stages

Globular embryo

Colchicine

Diploidization

Haploid plant

Torpedo-shaped embryo

Doubled haploid or haplodiploid plant

Principle of creating a haplodiploid

However, at present, the targeted transfer of a single gene is based on the technique of backcrosses (Inset 3.3), which like all the other pedigree techniques is time-consuming and costly. It is here that precise transfer of a gene by transgenesis very quickly gained interest among breeders, who have high expectations from this method of genetic engineering despite the inconveniences involved, especially its public image.

In selection, the control of male sterility in production of hybrid seeds must be specially mentioned. The introduction of a coding sequence for synthesis of a ribonuclease in the tapetum cells leads to anomalies in development and maturation of pollen grains, which leads to male sterility in these plants.

Inset 3.3

Backcrosses and transgenesis

The backcross technique is a particular case of pedigree selection in which the generations obtained are successively crossed with the same parent. At least a dozen generations are required for the selection, stabilization, certification, and multiplication of seeds. The same variety is obtained around three to four times as quickly by gene transfer.

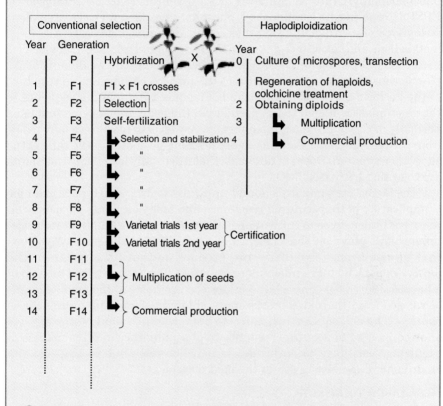

Comparative time required for variety obtained by backcrosses and by transgenesis

b) Acquisition of resistance

The acquisition of various types of resistance especially to phytosanitary products, parasites, or predators is by far the most important application of plant transgenesis. When the resistance is only partly acquired, it is called tolerance.

RESISTANCE TO HERBICIDES

Herbicides are molecules that destroy weeds and other "pest plants" while leaving the cultivated plants alone. The specificity of products is often low and the success is a result of the difference of sensitivity between two types of plant receiving a single type of chemical. Tolerance or resistance to herbicides is useful for the farmer because it allows easier, faster, and more discriminant chemical clearing. Resistance to systemic herbicides, such as glyphosate (or Round-up), has been the subject of two complementary strategies: weakening the sensitivity of the cultivated plant to the herbicide and developing in the plant a detoxification capacity, i.e., the capacity to degrade the molecule used. Glyphosate intervenes in the biosynthetic pathway of shikimates by competition with phosphoenolpyruvate at the level of an essential reaction catalysed by EPSPS (5 enolpyruvyl shikimate-3-phosphate synthase). The treated plant is deprived of its capacity to synthesize certain amino acids essential to its metabolism and growth (e.g., auxin, lignin, coumarin). The transfer of a bacterial gene coding for EPSPS into tobacco leads to the weakening of the plant's sensitivity to the product and the tobacco becomes more resistant to the herbicide. The resistance can be increased if a mailing sequence to the chloroplast is added to the protein via the gene; the enzyme functions normally in the chloroplast. A supplementary level of resistance may result from the positioning of the gene under a strong promoter in the cells of the shoot tip meristem. Doses of herbicide that are totally lethal for wild plants are tolerated by these GMOs.

The Monsanto company's Round-up-Ready soybean is a typical example of application of this acquired resistance and today accounts for nearly all the areas under soybean cultivation in the United States. Another herbicide, bromoxynil, plays on the presence of a nitrilase in the monocotyledons that allows them to detoxify the product and on its absence in the dicotyledons, which are much more sensitive. The transfer into tobacco of a bacterial nitrilase gene, placed under the promoter of the small subunit of rubisco (and therefore possessing a chloroplast mailing sequence), is expressed by an acquired resistance to bromoxynil. A similar strategy has allowed tobacco to acquire resistance to phosphinothricin. This process of acquiring resistance to herbicides is not, however, free of risks for the environment, as we will see in the next chapter.

RESISTANCE TO INSECTS

There are several strategies of transgenesis designed to protect cultivated plants from parasitic and predatory insects. One of them consists of reducing the populations of these insects by disturbing their moulting and thus their capacity to reproduce. It is known that, in order to moult, insects must secrete a moulting hormone, ecdysone, steroid in nature. For this, they need to borrow the sterols from the plants they feed on because they cannot themselves synthesize the precursors of the sterols. The sterol pathway is a well-known metabolic pathway in which the gene sequences that code for the synthesis of enzymes involved in all the stages are now identified and characterized. Researchers have been able to modify the sterol profile of plants by transgenesis and consequently reduce the usual development of their insect pests.

Another technique that is the source of great worry for our contemporaries is the Bt technique. For a long time the existence of a bacterial toxin secreted by *Bacillus thuringiensis* has been known and it has been used as an insecticide, particularly against certain butterfly larvae. The toxin modifies the permeability of the plasma membrane of intestinal cells of the caterpillar, disturbing its absorption of nutrition and the specificity of proton transfer linked to digestion. This leads more or less directly to the death of the insect. Bt is a biopesticide currently used in agriculture, silviculture, and horticulture. Its efficacy is limited for insects that dig galleries in the stems, such as maize pyralid. In 1985, Van Montagu and colleagues successfully transferred the gene that codes the synthesis of the toxin into tobacco (Tempé and Schell, 1987). The tobacco thus became resistant to the pyralid because the plant secreted its own insecticide. Various genes for toxins more or less specific to certain phytophagous parasites were isolated (by the Mycogene company) and transferred to cotton, tomato, and maize (Monsanto, Novartis). In cotton, the results were immediately encouraging, but resistance soon appeared among the predator populations. The application of the technique to maize is the best known and presently the best developed, especially in the United States and Argentina (Inset 3.4). Cultivation of Bt GMOs covers more than 90 million ha in North America and represents the majority of soybean and cotton crops and close to half of the maize crops. The commercial production of maize was met with hesitation and scepticism in France, where at first consumption was authorized but not cultivation, which caused a stir within the Commission de Génie Biomoléculaire. Subsequently, cultivation was authorized despite the protest campaigns of some anti-GMO organizations.

There are some alternatives to Bt genes that code for the synthesis of various other proteins with insecticidal effect, such as cholesterol oxidase or certain protease inhibitors. Also, plants have been transformed in order to make them produce molecules that attract specifically the proper predators of these phytophagous insects; these molecules often belong to the terpenoid pathway.

INSET 3.4

Creating a Bt maize

The gene for the toxin is identified and isolated from a genome of the soil bacterium *Bacillus thuringiensis*. This gene, placed under a strong promoter, is transferred by biolistics into cultured maize cells, from which fertile plants are regenerated. The plants express the Bt gene and are thus found to be protected from the attack of caterpillars and especially pyralid.

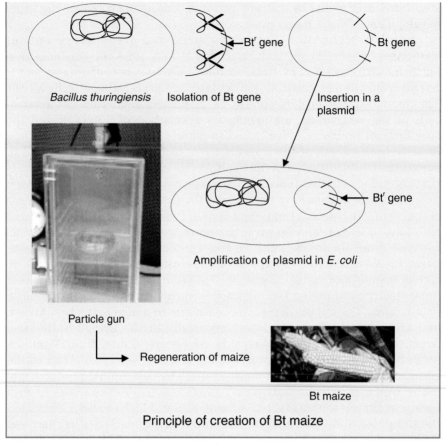

Principle of creation of Bt maize

RESISTANCE TO PATHOGENIC FUNGI

Cryptogamic diseases, i.e., those due to pathogenic fungi, that attack cultivated plants are particularly numerous and often severe. They may cause losses of 20 to 40% of yield depending on the place and the year. Some are still remembered as virtual calamities, such as potato mildew, which appeared in Ireland in the mid-19th century and caused a mass migration of the inhabitants of that country, especially to the United States. These diseases are controlled essentially by chemical antifungal treatments, which contribute significantly to contamination of ground water.

The search for a natural resistance, similar to what many wild plants possess in their genome, has oriented research towards programmes of gene transfer from this genetic reservoir represented by wild parents to their cultivated cousins. The desired genes can be transferred by hybridization followed by backcrosses as well as transgenesis. Here also, several strategies have been used. One strategy consists of introducing the gene coding the synthesis of a β-glucanase or a chitinase that has the property of attacking specifically the parietal structures of pathogenic fungi during their contact with the plant. A correct level of resistance sometimes requires the presence of several transgenes. In all cryptogamic attacks, three phases can be distinguished that can represent as many defence programmes: recognition of the pathogen by the plant, activation of an entire cascade of reactions in the plant cell, which puts not only itself on the defensive but also the neighbouring cells and even the entire plant, and finally the expression of defence mechanisms, especially by the secretion of molecules of the phytoalexin type (Inset 3.5).

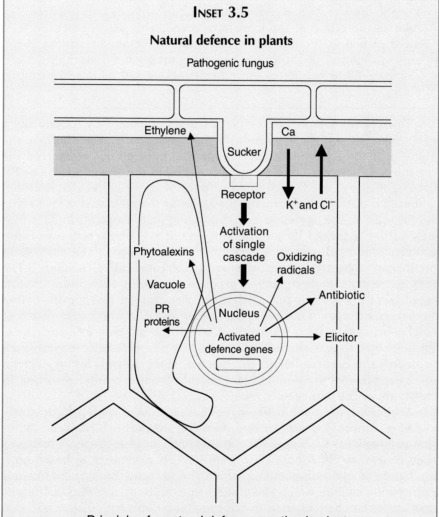

INSET 3.5

Natural defence in plants

Pathogenic fungus

Principle of a natural defence reaction in plants

When a predator, in this case a parasitic fungus, attacks the cells of a plant, especially if it inserts a sucker into them, a receptor membrane activated by an elicitor secreted by the fungus activates in turn an entire cascade of reactions involving ionic exchanges and the participation of G proteins, which transmit the information to the genome contained in the nucleus. In response, the cell activates its defence genes, which transcribe the messengers whose translation is the source of a multitude of defence substances—PR (pathogenesis-related) proteins, antibiotics (e.g., capsidiol), free radicals, and phytoalexins—all molecules that collaborate to protect the cell that has been attacked and the surrounding cells. If a gene for elicitor has been transferred into the plant cell, the activation of defence genes leads to the secretion of endogenous elicitor, which is thus a source of relative protection of the transformed cell. This set of reactions is accompanied by an emission of ethylene, which participates in the transmission of the information and alerts other geographically distant cells belonging to the plant itself as well as neighbouring plants.

The natural defences of the plant can be reinforced by introducing a gene coding for synthesis of an antifungal compound. For example, the transfer of the stilbene synthase gene of grapevine has reinforced the resistance of a tobacco that received the gene at the site of a rot. The PR (pathogenesis-related) proteins are part of these molecules secreted following the attack of a parasite; they present the peculiarity of not being very specific to the aggressor and thus being able to protect a plant that was earlier attacked by non-fungal parasites. The secretion of these protective molecules is often the direct consequence of the pathogen's production of highly aggressive protein molecules called "elicitors". The fungus *Phytophthora infestans* produces a clearly identified elicitor, cryptogenin, that triggers a defence reaction in the victim called "hypersensitivity". This reaction is expressed in the rapid death of neighbouring cells from the point at which the parasite attacked them. The pathogen thus works toward its own loss. The geneticist can only seek to amplify this natural defence mechanism. Thus, the gene for cryptogenin of *Phytophthora* has been transferred into tobacco plants, resulting in resistance to the pathogen.

At present, about 30 resistance genes capable of conferring a widely varied resistance (to fungi, viruses, bacteria) are known. It has been shown that potato plants transgenic for a glucose oxidase gene are more resistant than control to *Phytophthora infestans*. The resistance is based on a multitude of molecules among which the proteins have in common a richness in leucine. It is also linked to oxidation reactions. Different approaches can be combined to achieve greater efficacy. Research in this field seems to emphasize multifaceted protection rather than a narrow specificity.

RESISTANCE TO VIRUS

Viruses are the source of some plant diseases. For a long time it was believed that plants do not have an immune reaction and are thus susceptible to

INSET 3.6

Acquisition of virus resistance by transgenesis

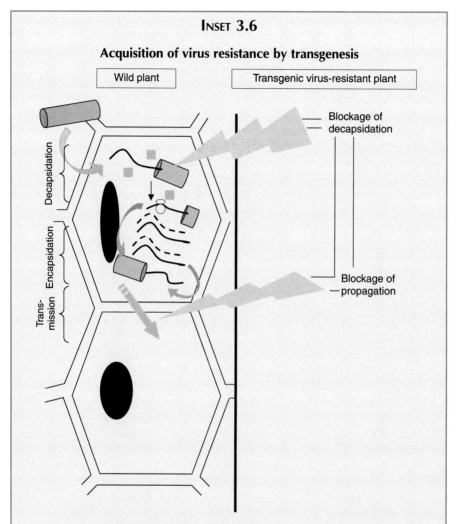

| Wild plant | Transgenic virus-resistant plant |

Decapsidation

Encapsidation

Trans-mission

Blockage of decapsidation

Blockage of propagation

Principle of acquisition of virus resistance by transgenesis

When a virus attacks a plant, it penetrates the peripheral cells, where it becomes "decapsidated", releasing a nucleic acid that supports its genome. This nucleic acid replicates and forces the host cell to synthesize new capsid proteins (or protein coats), which leads to multiplication of the virus. The viruses can then migrate to the neighbouring cells and thus spread the viral disease. In transgenic plants into whose genome a gene coding for the synthesis of a capsid protein has been inserted, the cells can no longer decapsidate the new virus, undoubtedly because of saturation of decapsidation sites. The virus can no longer multiply. Moreover, the migration of the virus towards neighbouring cells also seems to be disturbed. The disease cannot develop further.

virus attack. Ideas have evolved since then and we know that plants have a perception of self and non-self in some circumstances, especially in the case of reaction of pollen incompatibility during pollination. Researchers for a long time could only cure plants attacked by viroids, first by heat treatment and then by culture and regeneration of meristems *in vitro*. Nevertheless, these plants could be infected again. The method was thus sometimes only a provisional remedy and not a protection against viruses.

Conventional selection aims to identify varieties that are less sensitive or insensitive and then include them in crossing programmes. It is possible today to create virus-resistant GMOs (Inset 3.6). For an RNA virus, one method consists of isolating a fragment of viral RNA whose translation corresponds to a capsid protein. The inverse polymerase allows us to obtain a cDNA that is transferred into the host plant. Some plants, such as tobacco, potato, rice, cucumber, and pimento, have been transformed by means of cDNA and it was demonstrated that the transformed plants had become resistant to viruses. One of the explanations advanced is based on a saturation of decapsidation sites of host cells that deprives the cell of its capacity to decapsidate the viruses, which then invade the rest of the cell. Decapsidation is an essential stage for the replication of a virus inside the cell. Others propose the explanation of an incapacity of the transformed cells to recognize the virus.

This acquired resistance is called "pathogen derived" and has agronomic applications: the Asgrow company's zucchini, in 1996, and more recently melons produced by the French company Limagrain that were resistant to mosaic virus. The products of antiviral transgenes are sometimes called "plantibodies", by analogy with antibodies in animals.

It has also been observed that the acquisition of resistance to mycophytes could lead to a general resistance, including resistance to virus. It is not impossible, however, that there could be uncontrolled recombination between the virus and transgenes or between several viruses that parasitize the same cell, and caution seems to be necessary in this field of genetic engineering.

RESISTANCE TO PHYSICAL PARAMETERS

Several physical factors such as drought, low temperature, and salinity in the water supply have highly damaging effects on plant growth and reproduction. However, some wild plants tolerate extreme living conditions and accommodate themselves very well by structural or physiological adaptations. The plants may reduce their need for water by limiting the water losses caused by transpiration. Desiccation, as well as excess salinity or frost, activates genes that code for the synthesis of proteins called LEA, such as dehydrins, especially in developing seeds but also in the rest of the plant that is subjected to stress.

Researchers have created transgenic plants that overexpress LEA, especially rice carrying a gene of barley LEA, placed under the promoter of the rice gene. Such plants were more resistant to stress than the control.

Nevertheless, these LEA are not sufficient, and effective protection requires in addition the presence of osmolytes, osmoprotective substances of the mannitol or sorbitol type. Transformed tobacco overexpressing the bacterial gene of mannitol-1-phosphate dehydrogenase have an improved tolerance to salinity. The regulation of saccharose storage in cells clearly influences the quality of drought resistance. All the reactions of tolerance and resistance seem regulated for the most part by one hormone, abscisic acid, which plays a key role in perception of and reaction to osmotic stress. In case of dehydration, abscisic acid does not act directly on the activation of genes but in relation with transcription factors; certain genes that code for the transcription factors have been identified (CBF, DREB). Other genes have been isolated in plants that can tolerate long periods of drought and then overexpressed in *Arabidopsis* to give this transgenic plant an increased resistance to water loss. These fields of research are regarded as important especially in developing countries.

The process is nearly identical for improvement of frost resistance. It is known that plants can be acclimatized to cold by being subjected to progressive stresses, which activate a certain number of specific genes. Some of these genes have been identified through the discovery of mutants resistant to cold, particularly in *Arabidopsis* (*AtPLC1* gene coding for a phospholipase C). Other genes, such as *CBF1*, *CBG2*, and *CBF2*, have been identified as coding for the synthesis of proteins that are linked to certain sequences of promoters of frost-resistance genes. Finally, the synthesis of specific proteins that are said to be "antifrost", capable of reducing the freezing point, has been indicated in winter cereals (wheat, barley). These "antifrost" proteins are also effective against certain predators.

The fixed life of plants has led them to be highly imaginative in developing economical means of defence!

3.3.2. Growth and development of plants

a) *Hormonal equilibrium*

The development of a young plantlet is the result of the functioning of certain specific genes called "development genes", controlled by a range of hormones with complementary or antagonistic effects. Each new stage of development is characterized by a new hormonal equilibrium. Transgenesis thus has recourse to two strategies: to act on the genes themselves or to act on the control of their functioning through control of hormonal equilibrium.

With respect to genes for development, the process is still essentially cognitive. A large number of genes for development of vegetative parts are known at present, essentially because of mutant forms of *Arabidopsis*. These genes determine the identity, form, and number of organs especially in histogenic regions of the apical shoot meristem. The genes for floral organization are well known in *Arabidopsis* and *Antirrhinum* and the theories on flowering proposed by Coen and Meyerowitz (1991–1998) for these plants can be generalized to all the others. In this field, the demonstration is still far from being established because there are two particular floral types, one carrying two verticils of stamens and the other being that of an irregular flower. Transgenesis, even though it is well advanced, is still based on fundamental research, and economically justified applications are rare. They are, however, highly promising in horticulture, where the production of flowers with multiple perianth verticils could be favoured.

The strategy based on control of hormonal equilibrium is preferred in applications.

Plant hormones belong to two major metabolic pathways: auxin belongs to the shikimate route and most of the others, cytokinins, abscisic acid, and brassinosteroids, are derived from the isoprenoid route (Inset 3.7).

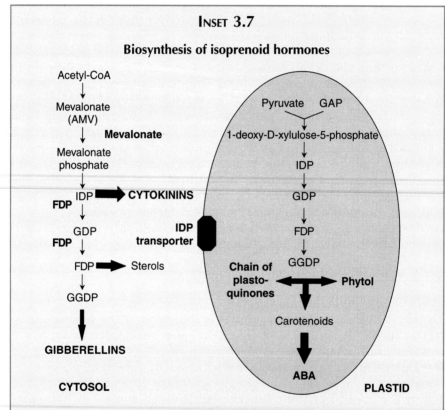

INSET 3.7

Biosynthesis of isoprenoid hormones

Biosynthetic pathway of isoprenoid hormones

In the plant cell, there is a double pathway of biosynthesis: the mevalonate pathway that leads, in the cytoplasm, to cytokinins, gibberellins, and phytosterols and the pathway of deoxy-xylulose-5-phosphate, which in the chloroplast results in carotenoids and abscisic acid. This double pathway of biosynthesis is characterized by successive condensation of molecules ranging from acetyl-CoA to phytosterols, passing through the crucial steps on which the pathway of cytokinin is branched (at the level of isopentenyl diphosphate) and that of gibberellins (at the level of farnesyl diphosphate). Amplification of the mevalonate kinase activity by insertion of multiple copies of the gene leads to an increased synthesis of cytokinins, whereas amplification of FDP synthase modifies the gibberellin/cytokinin equilibrium while increasing the pool of carotenes and certain phytosterols by the intermediary of an IDP transporter (S. Flajoulot-Champenoy).

A first generation of GM tobacco with modified hormonal equilibrium was created by introducing the *ipt* gene from Ti plasmid, but these plants had many malformations due to the continued presence of oncogenes. Genes isolated from yeast or other plants then took over.

Tobacco genetically transformed to overexpress mevalonate kinase and, consequently, capable of producing larger quantities of cytokinin have a considerably increased regeneration capacity and rate of development. This has ultimate effects on inflorescence size, seed production, and seed germination. Overexpression of farnesyl phosphate leads to a production three to four times as high of carotenes and phytosterols, giving the plant a greater resistance to illumination and to certain specific inhibitors of this pathway. It also leads to faster growth and greater vigour that researchers interpret as the availability of a larger pool of gibberellin molecules to the plant. However, this same pathway also leads to the synthesis of the antagonistic hormone abscisic acid and the instantaneous rapport of these two hormones must depend on fine control of genes involved in the pathway.

b) Nitrogen uptake of cultivated plants

The entry of atmospheric nitrogen into the biosphere is indispensable especially for the synthesis of proteins. However, few organisms directly convert nitrogen. Most plants take nitrogen from the soil nitrates by means of *nitrate reductase*, which by the intermediary of its mitochondria-chloroplast relationships allows the plant to assimilate nitrogen to synthesize amino acids. On the other hand, bacteria and cyanobacteria can directly assimilate molecular nitrogen of the air because of nitrogenase, an enzyme whose synthesis is coded for by the *nif* gene. Certain families of plants such as Fabaceae, some Mimosaceae, and some Betulaceae establish symbiosis at their roots with such bacteria and thus benefit indirectly from this uptake of atmospheric nitrogen. Procedures and assays have been proposed for grafting the *nif* gene directly in the genetic material of cultivated plants, particularly cereals, but it appears that it will be a few more years before they are successful. Meanwhile, transgenic tobaccos have been created with the gene coding for synthesis of nitrate reductase, which has indicated a close correlation between the level of expression of this gene and the nitrate levels contained in the leaf apparatus. Two applications follow from this development: the first is optimization of nitrate extraction by plants for their protein synthesis and consequently their growth, and the second is the use of nitrate-fixing macrophytes for decontamination of arable lands (spinach) or ground water (*Eichhornia*).

3.3.3. Control of seed germination and the "terminator" gene

The seed is the part of the plant that is often the easiest to produce commercially as a source of energy. This source can be used by animals and humans either as a source of organic matter and various directly assimilable products or as a source of future crops. Researchers from the very beginning sought to discover the principal genes that govern the essential steps of seed maturation and then its germination. Genetic control of most of the physiological parameters is now known, especially in the phase of grain filling and germination after maturation. An example of control of a gene regulating germination parameters is the gene called "terminator", over which much ink has been spilt because there has clearly been a huge gap between the motivations declared by its promoters and those imputed to them by the media and farmers. The promoters have said that this gene is designed to limit the risks of uncontrolled proliferation of the transgene by simple blockage of seed germination. The latter group have openly accused the proprietors of the technology of wishing to monopolize the transformed variety, contrary to the usage of the profession. Perhaps the truth lies between these two extremes. From the purely technological point of view, the strategy merits a close look.

It is a method of sterilizing plants by modification of the genome, resulting in high-yield plants whose seeds cannot germinate. If the technique is widely used, farmers will have no option but to buy their seeds from the companies that have patented them. One can easily imagine the total dependence of farmers that would result. The construction must be acknowledged as a beautiful feat of genetic engineering. Researchers have placed the desired gene under the control of an inducible promoter whose functioning depends on an external chemical signal. As long as the chemical is not applied, the promoter will not activate the desired gene. Thus, if this gene is lethal, as is the gene for barnase, a plant toxin, and the inducible promoter is active at the very beginning of development, the seeds will be destroyed by the endogenous toxin produced by the lethal gene from the first stages of embryo development. This process can be further perfected by a modification of the construction rendering the sterility of the plant conditional, so that a second chemical treatment can restore the fertility of the transgenic plant (see Inset 3.8).

The patent is held by seed companies such as Monsanto in the United States and Zeneca in England, and the detractors see in this technology a satanic construction, which they have named "terminator", designed to bind farmers to these companies for their seed supply. The companies that originated this idea are obviously agrochemical companies that control the sale of the restorative chemical product, whence the general outcry from associations for the conservation of biodiversity, politicians, governments, the Food and Agriculture Organisation, and even the UN General Assembly. In this affair, the explanation cited—protection against the dissemination of transgenes—has been ineffective in hiding the seed companies' more realistic motivation of a quicker return on their investment. Some see in this type of construction a dangerous weapon of agro-economic war.

INSET 3.8

Transgenic plant with controlled germination

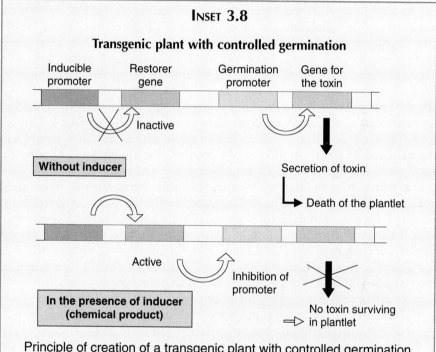

Principle of creation of a transgenic plant with controlled germination ("terminator" gene)

The creation of a transgenic plant with controlled germination involves two genes in tandem: the first is placed under a promoter that can be induced by a chemical product. This gene, called the "restorer", codes for the synthesis of a protein inhibiting the promoter of the second gene. The second gene codes for synthesis of a toxin that stops germination. In the absence of the inducer, the restorer gene is inactive and the promoter of the second gene, specific to seed germination, activates the functioning of the gene at the source of the toxin that is fatal to the plantlet. The seeds are thus not fertile. In the presence of the chemical product, the inducible promoter renders the restorer gene active. This inhibits the promoter of the second gene, which can no longer code for the synthesis of the toxin, and the plantlet, now viable, continues to grow.

3.3.4. Improvement of nutritional quality of food crops

Today there is one preoccupation of the agro-food industries that may be quite easy to justify in some cases, although those industries have received much criticism from consumers. Their objectives are generally qualitative, for example, a modified composition or greater starch content to improve digestibility. The frequent presence of anti-nutritional factors, especially in the legumes (soybean, lupin), has oriented conventional selection towards suppression, or at least drastic reduction, of such factors in plants grown for animal or human food. In this context, transgenesis has sometimes taken over from selection. The same is true of allergenic molecules, which are much more difficult to suppress effectively. Unlike anti-nutritional factors, which most consumers are sensitive to, allergenic products induce reactions in relatively few individuals or families or sometimes affect only some ethnic groups. Hence, industries are less interested in making a heavy investment, especially because of the great difficulty of understanding the mechanisms of allergic responses. There is no reaction specific to GMOs. The introduction of new food products in a population that is not yet accustomed to them frequently generates the appearance of allergic reactions. An often-cited example is kiwi fruit, which appeared in western markets in the second half of the 20th century and was the source of many allergic reactions when it was first sold. Soybean that was genetically modified for the production of a protein for nux vomica is a rare example of a GMO that was the source of an allergic reaction that ended in the death of a person, a fact that adversaries of GMOs never fail to cite.

There are many other examples of GMOs designed to improve the nutritional quality of food, but here we limit ourselves to two: vitamin A-enriched rice and fatty acids of oilseed crops.

Considering its importance in direct human food supply, rice is presently the focus of many transgenesis experiments. It is estimated that there are around 400 million people, distributed over more than 100 countries, who suffer from vitamin A and iron deficiencies. According to some studies, iron deficiency may be much more widespread. These deficiencies result in anaemia and blindness. Potrykus and colleagues, in Zurich, introduced in rice an entire series of genes that control the transformation of geranyl-geranyl di-phosphate into carotene, a precursor of vitamin A (Burkhardt et al., 1997). The promoter of these transgenes was chosen so as to activate the transcription of the gene only at the grain-formation stage. The grain takes on a pleasing yellow-saffron colour. Other genes inserted at the same time increase iron retention by the transformed plant while rendering it directly assimilable. This transgenic rice is no longer at the experimental stage and constructions have already been made in locally cultivated varieties to benefit the populations concerned.

Oilseed plants, such as sunflower, colza, peanut, olive, and, to a lesser extent, maize and soybean, produce large quantities of edible oil. Other plants, such as flax and *Camelina sativa* (gold of pleasure), produce industrial oils. A third category of plants, including castor and borage, are cultivated for pharmaceutical products. These oils, unlike mineral oils, are obviously biodegradable. The fatty acid composition, mainly oleic acid and linoleic acid, of oilseed plants has been the subject of many projects and patent applications. As of 1997, there were more than a thousand such patents.

Fatty acids are concentrated in the mature seed to serve as energy molecules for the embryo as well as to protect it against frost. Edible oils have 12 to 18 carbon atoms, while industrial oils have 20 to 24 (and waxes have even more). They also differ in the number of double linkages (generally 0 for saturated oils and 2 for unsaturated oils) and the position of such double linkages on the chain. The role of some enzymes called *desaturases* is to introduce double links in the chain. The genes for these desaturases have been isolated, cloned, and sequenced in *Arabidopsis*. Some have been transferred into oilseed plants, for example, to re-establish a correct ratio of α-linolenic acid, a fatty acid poorly represented in the oilseed plants, apart from borage, and the indispensable precursor to prostaglandin synthesis in humans. Oleate or oleic acid has uses in the agro-food as well as in the agro-industrial sector but, to be economically exploited, the oleic acid content of seeds must be close to 80% and the supply regular. In sunflower, selection methods were used to meet this requirement. Similarly, after years of rigorous selection, colza, which normally has less than 50% olein, also reached the requisite 80%. In this context, we can cite the example of a single operation of transgenesis on a haplodiploid that resulted in a colza with 76% oleic acid. We believe that without a moratorium on field trials, we would be able to achieve in just two or three years the kind of results that are obtained after several decades of pedigree selection.

The profiles of fatty acids of plant origin can be modified by transgenesis in oilseed plants. The coding sequence of the gene for a thioesterase in colza was modified to obtain seeds richer in stearic acid. A colza in which a gene from bay was introduced produced lauric acid.

In the fatty acid pathway, one of the difficulties of transgenesis arises from the fact that a large part of the metabolism occurs within organelles: plastids and mitochondria. Thus, we must find strategies to address the modification towards these organelles. For example, we can associate the gene sequence to the transit sequence of the gene of the small subunit of rubisco, which is known to penetrate this small enzymatic subunit in the chloroplast. Despite the difficulties of intervention at the genome level of these organelles, a certain number of results have been recorded to date. As a consequence of this process, there is a possibility of intervening in useful agronomic parameters such as frost resistance, which is known to be closely dependent on the lipid composition of tissues.

3.3.5. Horticulture and ornamental plants

Horticulture is an important economic sector comprising several products: garden flowers, cut flowers, perfumes, and vegetable dyes. These two last sectors have benefited from spin-offs of fundamental research on colours and fragrances of flowers.

New varieties of flowers are constantly created in horticulture. For a very long time the process involved intervarietal or interspecific hybridization. Later, horticulturists invested greatly in the research and production of mutants especially for the colour of floral pigments (e.g., *Gerbera*). For about 15 years, genetic engineering has come to complement the range of technologies available to the horticulturist.

Flower colour involves some of the most complex pathways of biosynthesis, and control over them is still only partial. It is often the result of a fine combination of original compounds and highly varied locations: green from chlorophyll, yellow from carotenoids, chalcones, and flavonoids, red or blue from anthocyanins. Some of these pigments are lipophilic and are located in the plastids. Others are hydrophilic and are located in the vacuoles. The pH of the compartment plays an important role and may vary according to the nature of the soil or the age of the plant. The absence of certain colours could result from the absence of the gene or from an insufficient transport between the compartments in which one of the precursors is synthesized and the compartment in which the finished product is stored. Blue is a rare colour because few plants possess the gene coding for synthesis of an enzyme of the 3-5 hydrolase type leading to the synthesis of delphinidine. Despite this complexity, some laboratories or companies have banked on the creation of new varieties by transgenesis. The first results were obtained in 1987 by the Florence company, which developed a brick red petunia expressing a maize gene. Since then, other genes have been transferred, such as petunia genes incorporated in a carnation that subsequently became mauve because it could produce delphinidine in greater quantities than its own pigments.

Fragrances are volatile compounds often derived from fatty acids, terpenoids, or phenylpropanoids. They play an essential role in attracting pollinating insects and consequently enhancing the reproductive capacity of the plant. The recognized chemical relationships between these substances and a certain number of phytoalexins suggest that fragrances could also have a role in plant protection. The synthesis of these compounds requires the presence of enzymes of the methyl-transferase and acetyl-transferase type: about six genes coding for these enzymes have been identified to date. These genes are particularly functional in the epidermal cells of flower organs and their regulation seems to depend on internal and external factors involved in their rate of transcription. Genes of floral fragrances of *Clarkia* have been successfully transferred into tobacco and petunia. Intraspecific transfers have also been done between varieties of rose and carnation so as to recover their original fragrances, which were lost during crosses designed to improve the appearance of the flower. The Nova Flora company, for example, transferred the gene for limonene syn-thase of mint into petunia. The flowers most susceptible to transformation are roses, carnations, petunias, chrysanthemums, orchids, tulips, gerbera, and gladioli.

The creation of varieties with new shapes, unusual colours, and original fragrances is the objective of many research laboratories. Curiously, agriculture could benefit by this same process but with a focus on reinforcing the capacity of plants to attract pollinators or improving the fruits in terms of commercial production.

3.3.6. Quality of industrial plants

In France, wood pulp is the third largest item in shortage in annual budget costs, after fuel and protein for animal feed. At the same time, France has the largest forest area in Europe and its forest heritage is greatly underexploited. Programmes have therefore been launched for the improvement of forest woody species reserved for cardboard or paper manufacture. The system of mandatory fallows has also helped reorient land planted with food crops towards industrial purposes. For example, lucerne crops have been planted not only to improve the soil nitrogen level without input of fertilizers but also for a multitude of industrial purposes, such as cattle feed, wood pulp for paper and cardboard manufacture, and the manufacture of certain plastifiers used in automobile accessories or electric insulators.

To satisfy the needs of all these sectors, it is necessary to exploit plants in which the lignin content would be very low (increased digestibility in feed, easier bleaching in cellulose pulp). The level of lignin, which is derived from the biosynthetic pathway of phenylalanine, must be quantitatively the lowest possible for good digestibility of feed and totally eliminated by extraction in the fabrication of wood pulp. However, the solubility of lignin varies greatly according to the species and the developmental stage of the plant. Its extraction is always difficult and sometimes nearly impossible. Thus, research was oriented towards reduction of lignin level by conventional selection as well as by genetic engineering.

It appears possible to introduce a gene in the antisense position in order to "extend" the activity of cinnamyl alcohol dehydrogenase (CAD), an enzyme involved in the final steps of lignin synthesis. This transgenic construction, done in tobacco, yielded modified lignins that were easier to extract and required a smaller quantity of bleach products, making it possible to reduce the level of pollutants introduced into the environment. Identical results were obtained with the antisense gene for cinnamyl-CoA reductase (CCR). More spectacular results were obtained with the antisense strategy applied to the gene for O-methyl-transferase, which reduces the degree of methylation of lignins by precisely replacing the methyls with hydroxyls, an operation that significantly and reliably modifies the lignification of cell walls and thus of fibres. This strategy was particularly advanced in lucerne for improvement of animal feed and especially to reduce problems of appetence and flatulence as well as for possible uses of the crop in the network of "fibres of plant origin". Similarly, double-transformant poplars seem to promise a brilliant future in the fibre network. Programmes with the same objectives have also been set up for spruce and eucalyptus.

3.3.7. Pharmaceuticals and parapharmaceuticals

The use of wild plants from various places to maintain the proper functioning of the human body or to treat diseases is common to all civilizations. The knowledge and use of medicinal plants goes back to the beginning of time. We need only look at the number of studies, books, and periodicals that have appeared on this subject as well as the number of botanical gardens we see today dedicated to medicinal plants. Every sector of our economy relies on plants or substances that can be extracted from them, especially for medicine, pharmafood, fragrances, hygiene products, and products to keep our skin, teeth, and hair healthy. The search for new molecules for medical treatment is still centred on a more complete inventory of resources of world flora; it is presently estimated that we know only 5% of those resources.

Here lies a considerable opportunity and potential for pharmacy in the new millennium that justifies, more than ever, the cooperation of botanists, ethnobotanists, and pharmacobotanists who can inventory the flora of regions that are still little known or unexploited in terms of natural products with pharmacodynamic effects. From the wild flora, it is a small step to programmed cultivation of plants for pharmacy and parapharmacy. It is thus not surprising that new varieties were subsequently developed using conventional breeding techniques as well as the new technologies. The plant cell was very quickly recognized as an economical structure for the production of molecules, including recombinant molecules with a pharmaceutical use. *In vitro* culture of these plant cells has the advantage of producing in "fermenters", as bacteria do, molecules in significant quantities and in highly competitive economic conditions. For the production of protein molecules, the behaviour of genetically programmed plant cells is close enough to that of mammalian cells especially for realization of the processes of post-translation maturation such as glycosylation, cleavage of a peptide signal, folding, the creation of disulphide linkages, and extracellular excretion. Moreover, plant cells are preferred to animal cells because, apart from their greater aptitude for transformation, in plant cells it is not necessary to eliminate the DNA of more or less pathogenic viruses used for their transfection, as is done in animal cell cultures. Let us also note that pathogenicity of plant viruses has never been observed in animal cells.

The use of GM plants for the production of medicine is today well advanced. An interesting breakthrough was achieved in the field of oral vaccines. At first, mice were immunized against the endotoxin produced by *E. coli* by being fed potatoes expressing a subunit of the toxin. Then, there was the project to vaccinate populations against hepatitis B by causing production of the toxin in transformed banana plants. Animals consuming the fruit from these genetically transformed plants would acquire the desired immune protection. Molecules of primary importance have been produced by GM plants. A brief list is given in Inset 3.9.

INSET 3.9

Molecules of therapeutic value produced from transgenic plants

The synthesis of some of the molecules present in the following list necessitated the creation of several successive GMOs. This is especially the case with human haemoglobin produced in transgenic tobacco (see Inset 3.10).

Partial list of molecules of therapeutic value produced from transgenic plants

Molecule produced	Plant of origin	Therapeutic treatment
Dog gastric lipase gene	Colza, tobacco	Mucoviscidose, gastric inadequacy
Interferon γ	Tobacco	Antiviral
Growth factor	Tobacco	Reconstitution of skin
Human serum albumin	Tobacco	Blood products
Human haemoglobin	Tobacco	Transfusion
Leech hirudin gene	Colza	Anticoagulant
Human encephalin	Potato	Antidepressant neuromediators
Subunit of *E. coli* enterotoxin	Potato	Immunization of mouse
Rabies glycoprotein	Tomato	Vaccination

The globin subunits were produced by tobacco under promoter 35 S and targeted to the chloroplast compartment to associate there with the haem molecule. The same principle was used to synthesize immunoglobulins that comprise light chains and heavy chains associated with a factor of protection against the proteolysis. The final assembly was done by crossing several lines of transgenic tobacco, each of which carried one of the four transgenes coding for the synthesis of one of the elements. The recombinant immunoglobulin thus reconstituted in all its complexity was shown to be functional. Synthesis by this process allows us to overcome the dangers from infectious agents carried naturally by the animal cells.

INSET 3.10

Human haemoglobin

Spatial representation of haemoglobin

Haemoglobin is a complex molecular structure comprising four globins
(2α and 2β) and a haem.

Diagram of the haem of haemoglobin

The haem and haemoglobin can be produced by lucerne or transgenic
tobacco. The complex assembly of the molecule is achieved by crossing
individuals that each have received one of the monomeric components
of the molecular structure.

The recombinant proteins thus produced are very often physiologically
active except when the last stages of maturation differ slightly between
the two kingdoms, especially at the level of glycosylation. Sometimes, to
avoid this difficulty, the targeting signals must be modified or induction
signals must be added so that the synthesis of the product can be triggered

only at a precise stage of development or maturity of the plant (e.g., fruit or seed maturation). It may be more convenient to stop the maturation of the molecule before the plant cell has conferred on it a specificity incompatible with the expected biological activity of the synthesized molecule. Some researchers have attempted to modify the synthesis of glycans so that the plant realizes a glycosylation of the protein, which would be of the same nature as that normally realized by animal or human cells.

One obstacle usually cited to the development of these techniques is the small quantity of recombinant proteins that plant cells can make without suffering from this unaccustomed task. The product is not always excreted and sometimes it must be extracted from the tissues or organs of the plant in which it accumulates (e.g., leaves, roots, seeds) and then purified. This is the case with serum albumin or human haemoglobin, and the process must be optimized before these products can be commercially produced. Some people feel nevertheless that in the long term the cost price is around five times less than for the products derived from blood plasma, not counting the advantage of diversification of agricultural products. Great hopes are invested at present in the production of gastric lipase by GM tobacco and colza under the influence of the gene isolated from the dog genome; this may solve the problems of people suffering from mucoviscidosis, which is manifested by an inadequacy of pancreatic secretions essential to the absorption of fatty acids. The world demand for anticoagulants could be covered by the cultivation of some hectares of transgenic colza that could synthesize hirudin by means of a leech gene and accumulate it in their seeds.

3.3.8. GM plants and environmental management

Lastly, one sector of the economy in which GMOs could play a predominant role is environmental protection. This is a recent but fully active preoccupation because of the increasing public sensitivity and the ever-increasing, widespread, and disturbing damage that human activities have done to our natural environment. Such protection could be preventive (avoiding pollution), curative, or remedial. In this last case, biological agents charged with decontamination of polluted areas could be bacteria, fungi, and lichens, as well as plants. Lichens and plants could also serve as *bioindicators* and be included in monitoring systems or stations. Certain administrative structures in charge of monitoring pollution have adopted tobacco as a major agent: the necrotic lesions on leaves can be monitored to judge the degree of pollution of an environment. Other plants are used for their capacity to fix and accumulate heavy metals and other undesirable elements. Some plants are known to be apt to concentrate certain metals, e.g., *Thlaspi caerulescens* for zinc or *Allyssums* for nickel, but they do not spontaneously constitute large biomass, so cultivated plants must be used in which fixing capacity has been amplified by genetic manipulation. Some types of cabbage and spinach have been programmed to accumulate ammonium, mercury, caesium, strontium, copper, or arsenic. Interesting results were obtained experimentally with *Arabidopsis* plants carrying a bacterial transgene that allowed them to effectively extract selenium. Plants generally react by accelerated translocation of elements extracted by the roots towards the aerial organs. At the cellular level, the element is detoxified by discharge and then sequestration in the vacuole. Metallic elements are fixed by a chelation reaction with various molecules such as organic acids or amino acids (histidine). If the accumulation is high, these plants can then be treated as an ore from which the accumulated product, after calcination and reduction to ash, can be extracted. These are the directions being explored by certain research centres of the Atomic Energy Commission. Environmental management is also a new sector of economic activity in which the presence of GMOs may not draw criticism from the public as it has in the context of food crops.

Chapter **4**

GMOs: Concerns and Remedies

4.1. QUESTIONS RAISED BY THE EXISTENCE OF GMOs

When we tally up the contributions of GMOs to our fundamental knowledge as well as our technical advances in their many applications over the past fifty years, we can only be optimistic, eager to go further, and relatively confident in the development of genetic engineering and its applications for the years to come. However, a large shadow looms over this field today: the public's nearly unanimous rejection of all the agronomic applications of these scientific studies, and the labelling of these studies as disastrous, dangerous, and hegemonic projects.

Such a manifest and even violent divorce between the scientific community and the public has rarely occurred in history. We must go back to the time of the development of nuclear energy to find an equally impassioned climate. We must also remember that non-military uses for nuclear energy were preceded by a tragic military past: bombs, deaths, fires, and abnormal births and cancers. Here, the damage has been nothing, or nearly nothing. And yet, questions have been raised about possible death linked to an allergic reaction following the absorption of a product from a transgenic construction. Should we not keep in mind all the benefits expected from gene therapy against the one or few deaths attributed, not proven irrefutably, to trials that were in their infancy, pioneering, or conducted without the necessary minimal caution?

Why the general distrust or rejection of this technology?

Several explanations have been advanced, and no doubt each of them is true to some extent. Let us look at the major objections currently made and the arguments that can be presented in response to them.

4.1.1. We have not had enough time to evaluate GMOs

The very short history of GMOs is an irrefutable factor that scientists must take into account in the establishment and execution of their projects. If we pursue our parallel with nuclear research, we will observe that between the discovery of radioactivity by Becquerel in 1896 and the first small power station, Zoe, in the Chatillon fort near Paris, around half a century had passed by. Nearly three-quarters of a century separated this discovery from the decision of nearly every French Electricity Board plant to go nuclear. In comparison, only about ten years have passed between the first transgenic plant produced in the laboratory and the first mature transformed tomato for human consumption. Similarly, only a few years separated the first experimental farm fields planted with GMOs and the 45 million ha of transgenic soybean cultivated in the world in the last years of the 20th century.

To respond to this criticism, the experimenter must demonstrate, with the techniques available now, the perfect harmlessness of the gene after it is transferred into a new genomic environment. The task is arduous and the need for it may seem insufficient with respect to the slight, but possibly long-term, effects of the transgene on the nature and the regulation of metabolism in the transformed plant. This quest can only be enriched with time, as the criticism is a sustained one. Conscientious and prudent scientists, although they do not deserve a prohibition that would handicap their search for knowledge, would abstain from precipitation in fields of possible applications and would be able to join their voices with those of their critics when they become aware that the requisite harmlessness has not been ensured. However, unlike the resolute and absolute opponents of all field experimentation, scientists can only insist on the need for a regulated experimental programme.

4.1.2. GMOs have arrived at an inopportune time

It is true that GMOs have developed during an awkward period in agro-food research, which has already suffered from many scandals. After chicken with high dioxin levels, consumers were faced with herbivores fed on meal of animal origin containing prion and responsible for bovine spongiform encephalopathy (BSE or "mad cow disease") and perhaps also degenerative diseases of the human nervous system, or on meal enriched with residues from sewage treatment plants. The recent past and concerns that are still a reality have tarnished the image of an industry that believed itself to be self-sufficient and capable of producing quality food. Older scandals that are still remembered and have nothing in common with GM plants are the source of confusions that are more tragic than comic, such as blood contaminated with the AIDS virus. These confusions, imagined by a few ignorant or ill-intentioned people, have become widespread, helped by gaffes committed by a press and media in quest of an audience. In this respect, the scientist cannot be held responsible for the time at which the biological sciences working toward mastering the genome have matured.

Any generalization is out of place, since these are distinct problems with nothing in common except that they leave the consumer worried, suspicious, and sometimes completely disoriented. Nevertheless, these unfortunate confusions have the merit of attracting the attention of consumers to the quality of their food, from the "pitchfork to the dinner fork". What has followed is a demand for and a process of traceability and treatment that have favoured new forms of agriculture such as organic farming or at least "rational farming" and, moreover, the establishment of labelled norms of quality in agro-foods.

4.1.3. GMOs and their derivatives are difficult to detect

Recall that GMOs must be distinguished from the products extracted from them, which are not considered GMOs and are not subject to the same laws or regulations. It is true that one of the most delicate problems associated with GMOs is their detection. Agro-food professionals must label each product as long as any of its components contains more than 1% of a GMO. That means we need to establish laboratories or organizations capable of detecting the possible presence of a GMO and measuring the level of it. Various protocols have been tried out and presently the most practical and efficient is to look for the trace of exogenous DNA that has been transferred and to estimate the possible number of copies compared to a given quantity of DNA. This search is done through amplification of DNA sequences by PCR using specific primers. First, DNA is extracted from a sample of the raw product (e.g., seeds or a piece of an organ). It is also possible to start with extracted products or those that have already been processed in some way (e.g., soyabean lecithin, cereal starch, tomato pulp, sometimes even a frozen dinner). Still, we must know what transgene we are looking for and have its sequence available in order to be able to "construct" the primers.

This seems simple when we have an idea of the nature of the transgene (for example, Bt gene in maize grains). But it is an entirely different matter when there is no such indication. In that case we could start by looking for genes frequently used in transgenesis, such as kanamycin resistance gene, 35 S promoter of CaMV, the NOS terminator, or the borders of T-DNA, but the absence of these sequences does not constitute a proof of the absence of GMOs. Thus, we must pursue the analysis by looking for genes that, for a given plant, have been the subject of a request for authorization from the Commission du Génie Biomoléculaire (CGB) or other authority. However, the number of these genes increases every year and there are many that have never been the subject of a declaration and may come from countries in which the legislation is much less strict. There may also have been a simple transfer of a certain number of supplementary copies of a gene placed under its own promoter and introduced in a plant by biolistics or liposome fusion. In that case there is no detectable element, although the plant is a GMO.

In addition to these qualitative tests, there are presently quantitative tests that involve parameters of much more rigorous samples. To determine the number of copies, for example, we must take into consideration the environment upstream and downstream of each copy because the insertions are done at different places in the DNA molecule. Quantitative tests are far from reliable and much more costly. Norms must be established by a standards commission. A certain number of small start-up companies are specialists in this sort of service (e.g., Clermont-Ferrand, Nantes, in France), but there is at present no certification that can be examined and the expertise of such companies can only be assumed.

The problem of positive detection of GMOs is far from being resolved and the certification of the absence of GMOs is still based on the product's traceability and the good faith and honesty of all the producers involved.

4.1.4. The risk of transmission of the transgene by ingestion

Transmission of the transgene by ingestion is a risk very frequently feared with, it seems, a corollary that is not always clearly expressed but that implies a possible integration of the transgene into the genetic make-up of the consumer. The transformant power of DNA through consumption of the molecule along with our food has never been observed or demonstrated, but then neither has its impossibility or improbability. All the food we ingest must pass through digestive juices and the hazards of pH, which renders any fear on this ground entirely unfounded, no matter what the food regime of the consumer. Cooking makes no difference: herbivores are not at risk of becoming grasses any more than a man can become a rabbit by eating one. There is no information, to our knowledge, in support of such a fear.

However, the presence of a new gene acquired by transfer in a traditionally edible plant is not to be treated entirely without an elementary principle of caution. The absolute harmlessness of the transgenic plant ought to be verified, considering that the gene, even if its function has been well established, may in its new genomic environment express itself by a synthesis of one or several unpredictable toxic substances. There could be interference with other genes because of the variability of insertion sites and the phenomena of co-suppression, disruption, or amplification that could result. Such interference could be expressed especially by the appearance of more or less serious allergic reactions without implying, in any case, a transmission of the transgene itself.

Over the long term, the risks to human health from frequent or regular consumption of GMOs are entirely unknown. This is why a research programme was established in 1999 to evaluate the alimentary risks linked to GMOs especially through animal feed (GMOCARE programme).

4.1.5. The risk of dispersal of transgenes

The control of possible dispersal of transgenes poses a genuine problem. In other words, there may be a danger that the transgenes introduced into one species or variety of plants could be found, against the intention of scientists, in other species or varieties with which it could cross sexually. The uncertainties in this field are considerable and only experience can produce answers. Certainly there are some plants, regions, and conditions of cultivation that make this phenomenon unlikely. For example, in much of Europe, maize has practically no chance of hybridizing with any plant of the local wild flora. Thus, there can be no uncontrolled gene flow. The only possible danger might come from an exchange of pollen between a GM population and a non-GM population. To prevent such an eventuality, the cultivation of GM maize varieties must be duly declared, subjected to control, and planted at a minimal distance from other fields (200 m), so that transgenic pollen cannot come into competition with other pollen. It has been advised that a buffer zone be planted around the field, made up of several rows of non-GM plants that could serve to trap wind-borne or insect-borne pollen. There are, however, people who consider this mechanism entirely inadequate, arguing that the minimal distance imposed is clearly insufficient because vector insects can travel much longer distances. The situation in Central or North America would be different from that in South America, where maize originated and where related species are still found (e.g., teosinte).

In Europe, the cultivation of colza presents an entirely different situation. There is a natural hybrid (*Brassica napus*) of two cabbages (*B. campestris* and *B. oleracea*) belonging to the family Brassicaceae (e.g., crucifers). It is a family widely represented in the wild flora of Europe. The cabbages can cross with mustards and rapeseeds and, in addition, the "double zero" colza that are presently cultivated have mitochondria of radish and chloroplasts of rape. Thus, a colza transformed to resist herbicides could transmit this resistance to precisely those weeds that are the target of the herbicide treatment. The result would be totally the inverse of what is counted on. This predictable dispersal of transgenes was the reason for a moratorium, at first for two years and subsequently extended. Some researchers and breeders (P. Guy, 1998) feel that cultivation of transgenic colza must be definitively abandoned. Others wish to evaluate the significance of transgene flow and for this purpose have set up field trials in perfectly controlled parcels of land. However, members of protest movements have chosen to destroy these field trials illegally; unfortunately, the violence with they have flattened and trampled on the crops has certainly defeated their purposes and helped to disperse the pollen!

It is thus necessary, before transgenic crops are planted, to delimit the cultivation zones clearly and to survey the precise nature of their environment. The same already applies for ordinary crops, which are necessarily affected by the type of agriculture practised in neighbouring fields. These precautions must be adapted for each crop, taking into account its characteristics of cultivation, reproduction, and capacity to hybridize. Perhaps we will arrive at a delimitation of areas of highly variable size for each type of crop, but the idea could seem constraining for a profession in which one of the ultimate attractions is the liberty of initiative. In France, laboratories such as those of P.H. Gouyon at Orsay or M.C. Chevre at INRA in Rennes have invested in research on the environmental impact of GM crops. While not underestimating the possible dangers, we must note that these plants, by producing one or several specific proteins, generally do not become more competitive and even suffer a "selective handicap" in relation to the wild plants, which are better adapted to the given environment. Exceptions are, of course, always possible. Moreover, one can also imagine an indirect impact, for example, through root secretions that could cause an imbalance in the fungal flora and modify the selection pressure established over time. This impact could be minimized by using only inducible promoters that function in precise circumstances such as a stress caused by a parasite or predator or only in some organs such as the leaves or flowers during the reproductive period. GMOs will thus have a more significant impact on cultivation practices than on the environment.

4.1.6. Genes for antibiotic resistance

Among the marker genes that allow us to test truly transgenic plants, and among all those that have escaped the transmission of the transgene, the genes for antibiotic resistance are the most often used. This practice does cause concern among specialists of the rhizosphere and ecologists who easily imagine exchanges and transfers of antibiotic resistance to soil bacteria and consequent modifications that could affect the equilibrium of soil microflora. There is certainly a ground for concern here, but it is an exaggeration to announce, as has been done time and again, that people could become resistant to antibiotics. As if they are not already so. As we can observe, some people do not hesitate to challenge even the fundamentals of medicine according to the principles of Pasteur.

It must be remembered that the antibiotics used for plant selection are often very different from those used against bacteria pathogenic to humans and animals and, consequently, the resistance genes are also very different. Research has been done on the impact of cultivation of transgenic plants and GM crop residues on the bacterial soil microflora. It has not successfully indicated an actual effect of these plants on the acquisition of resistance by soil bacteria: the frequencies of appearance of resistance are nearly identical between the control soil and the treated soil and are entirely of the same order as those of the appearance of resistance by random mutation. However, it has not been possible to demonstrate the total harmlessness of the use of antibiotic resistance genes, and wisdom dictates that we follow a principle of caution and consequently that the sequences conferring antibiotic resistance be withdrawn from GMOs after having contributed to *in vitro* selection operations. Such sequences can be withdrawn using tools of molecular biology such as restriction enzymes but the operation is often very delicate.

There are still only a small number of alternative methods available to bypass the use of an antibiotic resistance gene. There are, of course, colorimetric tests, such as the GUS test, which uses the gene for β-glucuronidase coming from *E. coli*; the enzyme coded for by this gene transforms its substrate (5 brome-4-chloro-3-indolyl-D-glucuronide) into a blue reaction product. Other researchers use a particular technique: the desired genes and the marker gene are introduced, not in tandem but independently on two distinct DNA fragments, and, among the descendants, those that have lost the selection gene but have conserved the desired gene are crossed and then selected. This is a delicate and time-consuming method. We have ourselves proposed an original system that consists of using a gene normally present in the plant but the overexpression of which, following the introduction of supplementary copies, clearly amplifies the regeneration capacity. When the culture medium is carefully selected, only those plants regenerate that are transformed by this gene, which is itself linked to the gene of agronomic value. An assay by PCR for the presence of the desired gene reinforces the reliability of the selection system. This procedure, perfected on tobacco, is still to be adapted to other cultivated plants.

There are some techniques that have never been used. The problem of antibiotic resistance is always topical and requires new research. The concerns expressed by consumers have urged legislators to propose a regulatory and judicial framework for GMOs. As very often happens in this kind of mechanism, technological advances are so rapid that the texts must be constantly updated and, correspondingly, there is some judicial licence that is taken advantage of by adventurers in GMO use, to the great regret of the public.

4.1.7. Inefficacy of transgenes in the face of appearance of resistance

We have seen that it is possible to make plants resistant to certain predators. However, we must ask whether that resistance proves lasting or, as is more probable, is temporary, particularly because of the possible appearance of phenomena of resistance in the parasite. It is true that the acquisition of a new property in a GMO as compared to the usually cultivated plants will exert a new selection pressure on the predator population, favouring the appearance of resistant predators that will then proliferate. It seems that resistance to pyralid has already begun to appear, particularly in cotton crops, and that the immense hopes invested in GMOs in this respect must be revised. To remedy the problem, some researchers advise the maintenance of a belt of fields planted with sensitive plants around the field planted with GM crops, a belt in which insect populations can proliferate that will in turn exert selection pressure on the resistant insect populations that appear. Moreover, some point out that the appearance of resistance may save some populations of butterfly (such as the monarch butterfly) that would otherwise succumb to insecticide secreted by the GMO. But in traditional crops, which are always subjected to repeated exogenous insecticide treatments, sometimes by aerial sprays, where is the concern for these poor butterflies? This aspect of the problem goes far beyond that posed by GMOs.

4.1.8. Multinationals and farmers

The creation of GMOs for the purpose of knowledge is a technology known in many public research laboratories: in France, there are INRA, CNRS, universities and other specialized agronomic institutions. In research in the private sector, genetic engineering and its agricultural applications are presently in the hands of large seed companies, national and international, which have at their disposal a network of satellite companies that covers many countries. This situation has disturbed a large number of farmers, because of the high prices at which seeds for GMOs will be offered to them, the impossibility of producing their own seeds, as they have frequently done in the past with certain self-fertilizing plants, and the royalties they will be subjected to if they use the species listed in the catalogue. The breeders who obtain the patents are still free to use the listed plants in programmes for crossing and improving plant varieties. The creation of GMOs, which has involved a high investment of genetic or genome research, has very naturally led certain companies, which desire a rapid return on investment, to file for patents. Thus, fresh questions are posed, with some urgency. Can genes be patented when international programmes funded by the public are supposed to decode genomes or make the knowledge acquired by this sequencing accessible and freely available? To whom do transgenic plants belong? These are subjects that periodically agitate the agricultural world, which has earlier encountered patents only in the context of farm machinery. The acquisition of a patent raises sticky problems of access to the diversity of genetic resources and encroaches on the freedom of choice of production. This freedom of choice is further threatened when the plant resistant to a product and the exclusive distribution of that product are both in the same hands. It is not an illusory problem because such ownership concentrated in a few large companies gives them the power over technological choices and the costs of risks that could be brought on a society and reinforces their position with respect to farmers in developed as well as developing countries.

4.1.9. Can GMOs be created clandestinely?

The clandestine production of GMOs is a problem that is rarely raised, but it is real. If we take up the parallel already drawn with nuclear energy, the difference is considerable. It is difficult to hide a nuclear plant in the countryside, while a "workshop" for the creation of GMOs could be as discreet as a press to print counterfeit notes. All that is required is a site of about 20 m², with a laboratory table, an autoclave (or even a large pressure cooker), a thermocycler, a small sterile hood, a small refrigerated centrifuge, pipettes, a refrigerator, some molecular biology kits—and an excellent scientific education! The minimal investment could be a few thousand euros and a comfortable budget would be 100,000 euros. The advocates of an absolute moratorium would do well to consider the possibility. It is for this reason that many scientists prefer a freedom under surveillance, with an obligation to account to the public, to a prohibition that would lead the more determined researchers toward clandestine research. The pressure exerted by anti-GMO movements is not without danger in some respects. Clandestine research is not the way in which large seed companies attempt to bypass regulations, but they could choose to locate their research activities in countries that are more tolerant or that respect the freedom of enterprise. Young researchers within the country would suffer the costs of such relocation. Furthermore, the production process would no longer protect people from a return of the transgenic products thus created to home markets and, no doubt to complicate attempts to detect them, mixed with non-GMO products. This scenario already seems to have been seriously considered by several multinationals and some companies may even have carried it out.

4.2. RESPONSE OF SOCIETY: REGULATORY AND JUDICIAL FRAMEWORK FOR GMOs

To the questions and concerns of citizens, society must propose solutions, a sort of code of good conduct, by setting up competent structures capable of explaining to the jurist and advising the legislator and by proposing an arsenal of texts, regulations, and laws that will unambiguously frame the creation, use, and sale of GMOs. It was for this purpose that some commissions were set up that, even though consultative, have through their opinions and decisions a determining influence on the policies of protective ministries. These bodies are expected to prescribe cautionary orders, plan controls, define rules for use and sale, delimit fields of authorization and prohibition, and, on a pan-European scale, to codify processes and streamline procedures.

   In France, two ministries are concerned with genetic engineering and decided to set up *ad hoc* commissions, the Commission du Génie Génétique or CGG for the Ministry of Research and the Commission du Génie Biomoléculaire or CGB for the Ministry of Agriculture and Fisheries, with the participation of the Ministry of Environment and Land Development.

These two commissions work independently of each other, even though some members may belong to both. They complement each other but have fields of intervention that sometimes overlap. For example, a university research laboratory must file an application with the CGG to get permission to experiment with and possess a GMO and file an application with the CGB if it is using an experimental greenhouse set up on packed earth or gravel. If the greenhouse has a cement floor, it is within the jurisdiction of the CGG. The CGB application is required if the laboratory needs to conduct field experiments.

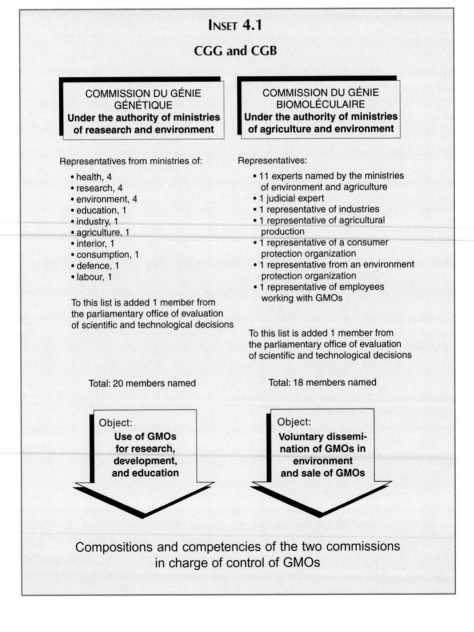

INSET **4.1**

CGG and CGB

COMMISSION DU GÉNIE GÉNÉTIQUE
Under the authority of ministries of reasearch and environment

COMMISSION DU GÉNIE BIOMOLÉCULAIRE
Under the authority of ministries of agriculture and environment

Representatives from ministries of:
- health, 4
- research, 4
- environment, 4
- education, 1
- industry, 1
- agriculture, 1
- interior, 1
- consumption, 1
- defence, 1
- labour, 1

To this list is added 1 member from the parliamentary office of evaluation of scientific and technological decisions

Representatives:
- 11 experts named by the ministries of environment and agriculture
- 1 judicial expert
- 1 representative of industries
- 1 representative of agricultural production
- 1 representative of a consumer protection organization
- 1 representative from an environment protection organization
- 1 representative of employees working with GMOs

To this list is added 1 member from the parliamentary office of evaluation of scientific and technological decisions

Total: 20 members named

Total: 18 members named

Object:
Use of GMOs for research, development, and education

Object:
Voluntary dissemination of GMOs in environment and sale of GMOs

Compositions and competencies of the two commissions in charge of control of GMOs

These commissions evaluate the demands from applications filed by the petitioner. The petitioner is heard at a hearing of the commission and may be accompanied by a person approved by the commission. Some days after the hearing, the commission communicates its decision to the petitioner, either approval or refusal of the request to conduct trials; in the latter case, reasons for the refusal are also given. The commission must then verify the application of its decision in the field but generally leaves this task to local plant protection services. When the trial is authorized, a document meant for the information of the public is posted in the town hall of the municipality in which the trial site is located, with the coordinates of the site, its area, and the nature of the plants tested (Inset 4.2). The field trial proceeds to the stage of growth planned or up to the harvest of grains. The plants are then burned and the soil is treated with broad-spectrum herbicides. The field is generally inspected in the year after the trial and is used for cultivation only much later. It is widely recognized that the commissions work seriously, competently, and conscientiously and that the trials are often proposed by well-informed and responsible persons who satisfy the criteria of the framework and of the strictest precautions.

However, these rules of transparency seem to have suffered somewhat since some protesters have taken on field experimental sites in a completely illegal manner.

Besides these instances of scientific evaluation and practical application, there are organizations for advice and orientation that can guide the regulatory and legislative procedures related to GMOs involved in food (in France, for example, the Association Française pour la Santé et la Sécurité Alimentaire). These organizations provide studies of GMOs, publish directives, and promulgate prohibitions and regulations.

The Convention on Biological Diversity discussed and adopted at the Rio Conference in 1992 was the basis of the establishment of a "biosecurity" protocol for the circulation and exchange of GMOs, designed to foresee the risks that GMOs may pose to the environment and human health. The products derived from GMOs were not discussed at that time.

The procedures, however, were very long and it was only at the beginning of the year 2000 that the negotiations ended in a compromise involving:

1. the obligation, for an exporter of GMOs, to obtain consent from the importer duly informed of the nature of the products;
2. the possibility, for each country, of prohibiting the import of GMOs if there are potential risks for conservation and sustainable use of biological diversity in that country or for human health.

These arrangements involve labelling, especially for seeds, proposed by directive 98/95 of 14 December 1998, enforced from 1 February 2000 onwards. The contents of the labelling, however, remain unclear. For example, for grains that are not used as seeds but that are transformed or processed, labelling is no longer required. In contrast, if the GMO is directly consumed (tomato, melon), the regulation (258/97) proposes a labelling according to specific modalities that were ultimately supplemented (1139/98) with respect to packaged derivative products destined for the end user. The same mechanisms pertain to additives and aromas obtained from GMOs. The labelling of derivative products is obligatory as soon as transformant DNA or proteins whose synthesis is directly linked to transcriptional activity of this DNA are present. For example, flour obtained from a transgenic maize must be appropriately labelled. If the absence of

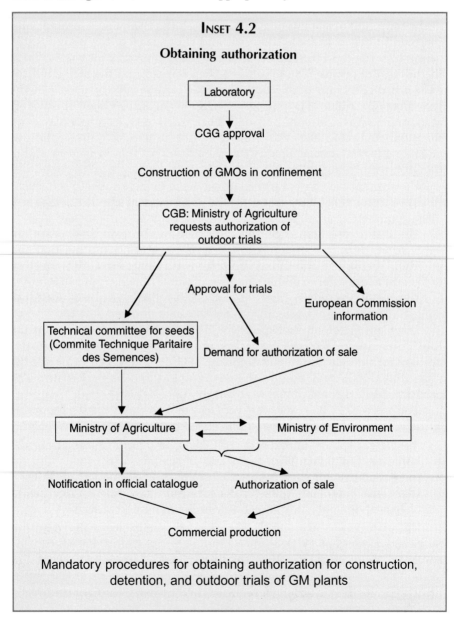

INSET 4.2

Obtaining authorization

Laboratory

CGG approval

Construction of GMOs in confinement

CGB: Ministry of Agriculture requests authorization of outdoor trials

Approval for trials

European Commission information

Technical committee for seeds (Commite Technique Paritaire des Semences)

Demand for authorization of sale

Ministry of Agriculture

Ministry of Environment

Notification in official catalogue

Authorization of sale

Commercial production

Mandatory procedures for obtaining authorization for construction, detention, and outdoor trials of GM plants

DNA and proteins is demonstrated, no labelling is required. Customarily, if the presence of the GMO is considered a chance contamination amounting to less than 1%, no labelling is required. As a corollary, the label "without GMO" cannot be indicated if there is such a trace. This is a strict but essential regulation.

The authenticity of the label, however, needs to be controlled in its applications. This control could be established in a laboratory, especially

---

INSET 4.3

Food products of transgenic origin

The table gives examples of food products of transgenic origin that have been the subject of a demand for authorization for sale, with dates of commercial production accorded and transfer to the member states of the European Community (source: Ministry of Economy and Finance, May 2000)

Company	Food product	Date of notification	Transmission of member states
AgrEvo	Oil from colza Topas 19/2	09/06/97	24/06/97
Plant Genetic System	Oil from colzas MS1, RF1, and the hybrid MS1 × RF1	10/06/97	24/06/97
Plant Genetic System	Oil from colzas MS1, RF2, and the hybrid MS1 × RF2	10/06/97	24/06/97
Monsanto	Oil from colza GT73	10/11/97	21/11/97
Monsanto	Ingredients* obtained from maize MON 810	10/12/97	05/02/98
AgrEvo	Ingredients* obtained from maize T25	12/01/98	06/02/98
Novartis	Ingredients* obtained from maize Bt11	30/01/98	23/10/98
Monsanto	Ingredients* obtained from maize MON 809	14/10/98	08/11/99
AgrEvo	Oil from colza Liberator L62	27/10/99	09/11/99
Plant Genetic System	Oil from colzas MS8, RF3 and the hybrid MS8 × RF3	27/10/99	09/11/99
AgrEvo	Oil from colza Falcon GS 40/90	27/10/99	

*Flour, gluten, semolina, starch, glucose, oil.

to verify the presence of the DNA and identify the raw material, ingredients, and finished products. The analysis must be accompanied by a process of traceability of raw material that allows us to identify their origin and their destination through all the stages of transformation and commercial production.

In France, genetically modified plants are presently authorized at two levels: either for any use—import, cultivation, transformation—or for import and transformation only. In the latter case, production within French territory is prohibited. In the first category are a herbicide-resistant tobacco and three maize varieties (Novartis, resistant to herbicide and pyralid; Monsanto, resistant to pyralid; and AgrEvo, resistant to a herbicide). All these crops are under the control of the CGB and biovigilance committee. The second category includes a soybean resistant to a herbicide (Monsanto) and a maize resistant to pyralid and to a herbicide (Novartis). The cultivation of these two last varieties is thus not authorized. This authorization also pertains to the cultivation of some transgenic flowers (Florigene). To this short list must be added some products authorized in countries of the European Union, the authorization of which is thus, in principle, valid for all the countries of the Union. This applies, for example, to a tomato of British origin (Zeneca) sold as puree. There are some exceptions, as with Topas colza 19/2 of the AgrEvo company, authorized in the European Union but included in the "colza moratorium" decreed in France. This is due to a safeguard clause promulgated by directive 90/220, but the decision of one member state must be approved or refused by the European Union within three months. Such prohibitive or restrictive measures have already been taken by several countries. The current prohibition on colza does not apply to oil extracted from it. For third countries, the imports are subjected to the usual national regulations (Inset 4.3).

Despite all these international directives and national regulations, a certain amount of imprecision still masks the traceability of products and activities of transformation, as well as the transparency of commercial transactions in agro-foods. A complete mechanism is still to be established for safety and fairness in labelling and monitoring food for human consumption. These advances in the translation of legislative texts are proposed in the law on agricultural orientation (99.504 of 9 July 1999). The texts must reinforce an understanding of the principle of caution and of all the networks infiltrated by the detection, transformation, and commercial production of GMOs. They must help streamline methods of detection and analysis in order to prevent litigation and commercial prejudices linked to inconsistent or insufficiently precise analytical techniques.

4.3. CONCLUSION: WHAT IS THE FUTURE OF GMOs?

Genetic transformation, or transgenesis, seems still unevenly mastered among the various groups of living things. Recognized as being highly advanced in bacteria and eukaryote organisms—particularly yeast—transgenesis in plants may be considered to have reached a certain maturity. It is certainly more advanced than in the animal kingdom, where GMOs are still at the experimental laboratory stage, particularly in man, where they are often only a subject of reflection. Immense advances, initiated and sustained by the search for solutions to problems of genetic diseases, are expected in the decades to come. We can expect some smoothing of the edges between the two kingdoms in time.

Genetically modified plants are certainly, at the beginning of this new millennium, at a decisive turning point in their history. They are still too recent for a calm and objective judgement to be made without risk, but already too advanced and present in our lives in the shape of food for the adventure to stop here, in simple indifference or misunderstanding. The period of discovery and the first implementations is already past and has given place to a period of development and perfection. The perception of all this must now allow us to better consider their destiny.

On the subject of GM plants, the countries of North and South America have seemed only slightly concerned although more than 90% of the GM plants cultivated in the world are in these continents. Europe, the cradle of GMOs, seems much more concerned although the areas planted with GMOs are still insignificant and although no real danger has yet clearly been established. The developing countries either express no opinion or declare themselves fervent defenders of GMOs. The encounters between researchers from different countries is also highly instructive. On the one hand there is the rather mocking laughter of American researchers who are shown the precautions taken in a transgenic greenhouse, which include airlocks between the various sections, filters on the openings, and an individual purification waterstation with chlorine injection. On the other there is the Egyptian agronomist who explains that the presence of GMOs, particularly Bt maize, concerns him much less than the potential benefits accrued from high yields that would not otherwise be possible on the 8% of Egyptian territory that can be cultivated to feed a population of 69 million.

In Europe, and particularly in France, the overall future of GM plants is highly uncertain and varies considerably with the field of application considered. Three possible attitudes must be considered, depending on whether we are looking at fundamental research, agronomic and agro-food applications, or applications in the field of medicine.

In fundamental research, it is unthinkable that we would do without a tool that has performed so well for a refined cognitive approach to the physiology and genetics of plants. At a time when *Arabidopsis* has been entirely sequenced (December 2000) and we are yet to find the precise function of thousands of genes that are sequenced, researchers will not renounce genetic transformation, one of the most reliable, effective, and quick methods available to them. On the contrary, there will be much more rapid development of GMOs of all kinds, "better" GMOs, free of selection sequences, if possible, in which the transgenes are generally expressed under the control of promoters that are inducible and thus can be perfectly controlled.

In the agronomic field, great difficulties can be foreseen for field cultivation of GMOs because of the suspicious and negative reception of a significant part of the population, well organized and informed by groups, movements, associations, and syndicates whose ideas are relayed by highly effective media. This orchestration did not arise without reason and we must recognize that many of the blunders must be laid at the door of the creators of GMOs and the companies that financed them. The difficulties will be much greater with respect to crops destined for direct human consumption. This position will be more or less that of most European countries, whether or not they belong to the EU, with probably surprising specificities and exceptions about this or that cultivated plant. We can foresee that there will certainly be some movements and convulsions in the Americas but transgenic crops there have been comfortably in place for some years. The commercial practices of those countries, which consist of exporting combinations of GMO and non-GMO products, will not simplify the information of European consumers.

With respect to the field of health, the production of GMOs, especially those that can contribute to phytopharmacology, will face much less difficulty because of the high stakes perceived by the public, who have been reassured by media that are apparently much more tolerant, even enthusiastic, about operations of transgenesis that directly affect humans than about those that affect our food. We can safely predict that transgenesis for therapeutic purposes will continue to develop, including in products for simple individual comfort. For example, it is astonishing to see the indifference and even the goodwill with which the public, the media, and the usual pressure groups have received the decision of certain cosmetic laboratories to launch promotion campaigns for a line of products based on "plant DNA", designed to tone the skin, when the manipulation of exogenous DNA is delicate and must be undertaken in a laboratory with an abundance of caution.

These attitudes, which appear often illogical and contradictory, reinforce the impression of insufficient distance to develop a maturity that would help us express a sane and objective judgement. To work toward such an objective judgement, political authorities in France have established a national committee of bioethics charged with proposing appropriate regulation and legislation. This committee can expect to work very hard and exercise a great deal of imagination and ethical debate in following the evolution of results and the appearance of new stakes and in constantly reviewing its position and proposing responses adapted to the concerns of citizens.

What will remain of these present debates between pro- and anti-GMO groups? Must we look forward to the victory of the latter, which are presently much more numerous? It is too early to tell, but it seems reasonable to consider that there will remain little more than the memory of passions revised and shaky convictions, or even the bitterness of lost opportunities. We cannot have a victory of one over the other because the two groups have contributed equally to the same progress. The former have promoted knowledge, found solutions to acute problems, and opened up new perspectives with still unknown applications. The latter have also made an important contribution by demanding a necessary prudence, requiring a rigorous framework for research, and imposing essential precautions and a constant monitoring of results. We all have to travel the same road, in the same direction, with its inevitable clashes and fruitful confrontations, like any "school" that has emerged only because of its cortege of supporters and critics.

So what is the future of GMOs? The first ten years of the 21st century will decide it. An absolute respect must be observed for the profound convictions and motivations of each side, calling constantly for prudent waiting periods and imbued with great tolerance, but firmly turned towards a future in which we must inevitably live with GMOs.

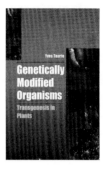

Glossary

Acetyl-CoA: Small molecule comprising an acetyl group linked to coenzyme A by an easily hydrolysed linkage. This molecule practically constitutes a metabolic crossroads within the cell and is at the basis of the large majority of metabolic routes: amino acids, shikimmate, isoprenoids, etc.

Agrobacterium: Gram-negative soil bacterium capable of parasitizing higher plants, particularly dicotyledons, and transferring to them part of the DNA of large plasmids that it contains. The best known is *Agrobacterium tumefaciens*, responsible for crown gall disease. It is widely used in plant transgenesis because part of the transfer DNA of its Ti plasmid can be replaced by DNA segments containing the genes to be transferred.

Allele: One of the forms of a gene. In diploid individuals, genes are present in the form of two alleles occupying the same locus on homologous chromosomes.

Antigen: Molecule that can be identified very specifically by an organism's immune system. An antigenic reaction in plants, for example, is the pollen-stigmata reaction during pollination preceding fertilization.

Antisense RNA: Complementary RNA of a transcript of a gene placed in the inverse direction of the native gene. It can thus hybridize with the RNA of the gene and prevent any exploitation by polymerases and particularly its translation into a protein.

Apical: Describes cells located at the tip of meristems.

Arabidopsis thaliana: Small plant of family Brassicaceae, widely used in molecular biology for the small size of its genome, its small number of chromosomes ($2n = 10$), its prolific reproduction, and its short vegetative cycle, which can be reduced to 2 or 3 months (4 to 5 generations a year).

ATP synthase: Enzyme complex whose basal part is incorporated in the internal membrane of mitochondria or thylakoids of the chloroplast. It catalyses the formation of ATP from ADP and of phosphate during respiration or photosynthesis.

Axenic (culture): *In vitro* culture that is free from external contaminants and internal symbionts. Also called "pure culture" or, incorrectly, "aseptic culture".

Base pairs or bp: Two nucleotide bases of a DNA molecule linked by hydrogen bonds.

Biolistics: Technique of direct genetic transformation using a gun with metallic microbullets on the surface of which DNA fragments are "stuck" and that are projected violently into the cells by the release of an inert gas, often pressurized helium.

Capsid: External coat of a virus formed by automatic assemblage of repetitive protein elements called protein coats; the capsid can take well-defined geometric form sometimes similar to the form of crystals.

Cell fusion: Modalities of union of two cells, often somatic, which combine their cytoplasm, their membranes, their organelles, and, in some cases, the contents of their nuclei.

Chaperon (molecule): A protein molecule participating in the migration or folding of another molecule.

Clone: Population of identical cells descended by successive divisions from a single cell without genetic modification.

Coenzyme A or CoA: Small molecule fixing on acyl groups (acetyl CoA).

Consensus sequence: Identical or partly identical DNA sequences belonging to genes that may have the same function or different functions.

Cosmid: A plasmid cloning vector for transfer of DNA segments that contains the two cos ends of bacteriophage viruses.

Crown gall: Disease caused by *Agrobacterium* and expressed by cellular proliferation and formation of a growth at the crown of the plant (stem-root junction). The *Agrobacterium* causes this disease by transfer of part of the DNA molecule from a large plasmid called Ti plasmid.

Cybrids: Organism derived from the fusion of two somatic cells that possess the entire cytoplasm of two parental cells but generally the nuclear genome of only one of the parents.

Cytokinins Family of molecules derived from the mevalonate route and involved in cell multiplication, growth, and development of plants.

Deletion: Loss of a DNA segment and the genes carried by the segment.

Diploid: Describes cells possessing all the homologous chromosome pairs and consequently the two alleles of a single gene.

DNA library: A set of fragments of DNA molecules cloned in bacterial plasmids and representing the entire genome of an individual (genome library) or all the complementary DNA of a genome, realized from messengers produced by cells at a stage at which the variety of messengers transcribed is the greatest, e.g., embryo stage (complementary DNA library).

DNA sequencing: Determination of the series of nucleotides in the DNA molecule.

Domain: Free region of a long DNA molecule, located between two regions fixed to the lamina, in the interior of the nucleus. Also used to describe a region of a protein that has some physiological autonomy.

Duplication: Doubling of a DNA segment and the genes carried by the segment.

Endonuclease: Enzyme that can hydrolyse the phosphodiester bonds of nucleic acids, single or double strand DNA or RNA.

Enhancer: Regulatory sequences of a gene capable, particularly, of stimulating the activity of a promoter.

Eukaryote: Cell of an evolved organism possessing a nucleus bounded by a membrane and cytoplasmic organelles.

Exon: Segment of the coding sequence of a gene that will be transcribed and maintained after maturation of the transcript and that has a specificity at the protein level. Opposite of intron, a messenger region eliminated in the course of maturation.

Exonucleases: Enzyme capable of hydrolysing the phosphodiester bonds towards the ends of the DNA molecules.

Genetic engineering: Form of human control of the genome of living things.

Genome: The total genetic information with a cell or organism. The maternal and paternal genomes form the genome of the embryo.

Genotype: Genetic constitution of a cell or individual, i.e., the potentialities of information in it. Differentiated from phenotype, which is the realization of those potentialities.

G protein: Heterodimeric protein linking to GTP and involved in the transmission of messages within the cell.

Haploid: Describes a cell or organism that has only half of the base number of chromosomes. This is the case with gametic cells, spores, and organisms derived from the development of these cells without fertilization (gametophytes).

Heterosis: Properties acquired and expressed by hybrids that can exceed the qualities of the two parents.

Heterozygote: Diploid cell possessing genes in the form of different alleles.

Homeobox: Common DNA sequence of around 180 bp present in a certain number of genes involved in development and expressed particularly during the growth of certain organs.

Homeotic genes: Genes having consensus sequences in their promoters that respond often to a single signal and are responsible for the development of organs during ontogenesis.

Homozygote: Diploid cells possessing genes in the form of identical alleles.

Hybridization: Pairing of two complementary nucleotide sequences: DNA/DNA; DNA/RNA; RNA/RNA.

Inbreeding: Depression or loss of vigour observed in self-fertilizing plants that can result from an accumulation of small mutations due to the drift of the population towards homozygosity.

Insert: DNA fragment transferred and inserted in a genome.

Insertion: Integration of a DNA segment in the DNA of a chromosome.

Intron: Non-coding region in the sequence of a gene. The introns are transcribed but excised from the RNA during its maturation. Thus, no manifestation is found at the protein level.

Inverse or reverse transcriptase: An enzyme isolated from retroviruses that synthesizes DNA from an RNA matrix.

Isoprenoid: Molecule belonging to a large family, an intermediary in the biosynthetic pathway of sterols whose base element is a unit of five carbon atoms: the isoprene.

Kanamycin: Antibiotic the resistance to which is frequently used for selection of transformed plants.

Kinase: Enzyme that transfers the phosphate group of ATP to a protein to phosphorylate it.

Ligase: Enzyme that serves to join or ligate two fragments of DNA molecules by the formation of a phosphodiester bond. This is a reaction that requires energy.

Liposome: Artificial structure composed of one or several lipid bilayers that can contain various substances, particularly DNA.

Microinjection: Injection of molecules, particularly DNA fragments, into a cell using a microneedle held by a micromanipulator.

Microsatellites: Very short sequences of DNA frequently repeated and located at parts of the genome specific to each genotype. Contribute to the establishment of a genome map.

Mutation: Spontaneous or induced modification in the nucleotide sequence of a gene that can be transmitted to the progeny of the cell.

Northern blot: Technique of transfer of RNA to a support membrane that can then be detected or identified using a specific probe.

Nucleosome: Basic structural element of chromatin constituting a "bead" that links a DNA sequence to a "core" formed of an octamer of histones.

Oncogene: Gene giving a cancerogenous trait to a cell.

Oosphere: Female gamete in plants. Not to be confused with the ovule, which is the organ that contains it.

Operon: Linked genes having a common metabolic activity and often a common regulation.

Peptide signal: Sequence of protein that determines its final location within the cell. It is particularly present to ensure the passage of the newly synthesized protein of cytosol into the internal cavity of the reticulum.

Plasmid: Small, circular DNA molecule that is independent of the principal genome of the bacterium, endowed with an autonomy of replication and often represented in several models. For example, Ti and Ri plasmids of *Agrobacterium*. These plasmids are very widely used for cloning of DNA molecules.

Pleiotropic: Describing genes involved in multiple metabolic routes and having several phenotypic expressions.

Polylinker: Particular region of a plasmid that allows mutiple cleavages by restriction enzymes and the insertion of an exogenous DNA.

Polymerase chain reaction (PCR): A technique of amplifying DNA fragments *in vitro* following alternate cycles of temperature variations in the presence of a particularly thermostable polymerase: the *Taq* polymerase.

Polyploid: Describes a cell that has several times the base chromosome number. Many cultivated plants are polyploid.

Polysaccharides: Saccharide polymers in linear or branched form, e.g., starch and glycogen.

Prenylation of proteins: Covalent link of an isoprene and a protein.

Primer: Short sequence of DNA or RNA that can be used, when paired with a DNA matrix, for synthesis of the same nucleic acid by elongation in the $5' \rightarrow 3'$ direction. A double primer is used in PCR reaction.

Probe: Fragment of a DNA or RNA molecule marked either with a radioactive element (radioactive probe) or by a chemical (enzyme) or physical element (fluorescence) and used *in vitro* to detect a complementary nucleotide sequence by hybridization.

Prokaryote: Organism lacking a nuclear compartment, as in bacteria, cyanobacteria, and mycoplasms.

Promoter: Nucleotide sequence that controls the transcription of a reading frame of a gene. This is where RNA polymerase is fixed.

Proteoglycane: Protein linked to one or several glycane chains (sugar).

Protoplast: Plant cell in which the cell wall has been digested by pectinases and cellulases.

Reading frame: Position of bases that are distributed in triplets so that each triplet can code for the fixation of an amino acid during translation.

Remediation: Rehabilitation of land by decontamination of the site, sometimes using plants that concentrate the pollutants.

Replication: Duplication of a DNA molecule.

Repressor: Protein that links to a regulatory region of DNA in order to block the processes of transcription of a gene located immediately downstream.

Restriction enzyme: Enzyme belonging to a family of nucleases and capable of cleaving the DNA molecule at highly specific places as a function of the succession of nucleotides present at the restriction site. The cleavages can be direct or delayed (and, in that case, sticky). These enzymes play a fundamental role in the control of the genome and the development of genetic engineering.

Restriction map: Diagram of a genome or part of a genome of an organism on which the various restriction sites are indicated.

Retrovirus: Virus with RNA but requiring the synthesis of an intermediary DNA for its replication.

Ribonuclease or RNase: Enzyme that serves to hydrolyse phosphodiester linkages of RNA.

Screening: Trial or selection of individuals within a population or a clone according to a specific criterion.

Selectable marker gene: Gene easily identified or selected that accompanies the target gene in tandem.

Somatic cell: Any cell of an organism other than reproductive cells or their close parents.

Southern blot: Protocol based on the separation of DNA fragments by electrophoresis and their specific identification by means of radioactive probes.

TATA box: DNA sequence generally located near the gene promoter in the eukaryote cells between −170 and −80 bp before the beginning of the coding sequence. This sequence is involved in the initiation of transcription.

Terminator: Sequence of a gene that gives the signal for the end of transcription.

Ti plasmid: Plasmid of *Agrobacterium* of which one part, T-DNA, is transferred in the plant cell ensuring its genetic transformation.

Transcript: RNA derived from the transcription of DNA.

Translation: Operation consisting of synthesizing a protein by means of a messenger RNA.

Transposon: DNA sequence that can move to the interior of a genome and can have genetically different reading or interpretation in its new environment.

Vector: Intermediary serving to transfer a gene from one cell into another. The vector may be a virus or a bacterial plasmid. A cloning vector is represented by the same elements but for the purpose of carrying a gene that is to be cloned. An expression vector is created so that the gene that it carries can be exploited by the transcription enzymes.

Vir genes: Genes of virulence that allow *Agrobacterium* to recognize specifically the cells of certain plant categories or families.

Western blot: Process of separation of proteins followed by identification by a probe (e.g., antibodies).

Yeast artificial chromosome (YAC): A cloning vector that behaves like a normal chromosome and follows the rules of cellular genetics. It is used to clone huge DNA fragments within yeast cells.

Zygote or egg cell: Cell resulting from the fertilization of a female gamete by a male gamete and therefore at least diploid.

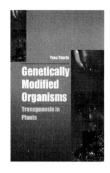

Bibliography

General References

Bernot, A. 2000. *L'Analyse des Génomes, Transcriptomes et Protéomes*, 3d ed. Dunod, Paris.

Campbell, A. 1995. *Biologie*. De Boeck, Brussels.

Cenatiempo, Y. and Julien, R. 1995. *Biotechnologies Aujourd'hui*. Pulim, Limoges.

Lodish, H., Darnell, Y. and Baltimore, D. 1997. *Biologie Moléculaire de la Cellule*. De Boeck, Brussels.

Robert, D. and Roland, J.C. 1998. *Organisation Cellulaire*. Doin, Paris.

Scriban, R. 1996. *Biotechnologie*. Lavoisier, Paris.

Smith, J.E. 1996. *Biotechnology*, 3d ed. Cambridge University Press.

Susuki, D.T., Lewontin, R.C. and Gelbart, W.M. 1997. *Introduction à l'Analyse Génétique*. De Boeck, Brussels.

Tourte, M. 1998. *Introduction à la Biologie Cellulaire*. Diderot, Paris.

Tourte, Y. 1998. *Génie Génétique et Biotechnologies: Concepts et Méthods*. Dunod, Paris. 2nd ed., 2002.

Watson, J.D., Gilman, M., Witkowski, J. and Zoller, M. 1997. *L'ADN Recombinant*. De Boeck, Brussels.

Widner, F. and Beffa, R. *Aide-Mémoire de Biochimie et de Biologie Moléculaire*. Lavoisier, Paris.

Popular journals

Genetically modified plants have been the subject of a great many articles in popular scientific journals of a high calibre. We cite the following in particular:

L'Année Biologique
Biofutur
Grandes Cultures
Nature

Pour la Science
Proceedings of the National Academy of Sciences (PNAS)
La Recherche
Science
Sciences et Avenir
Scientific American

Specialized References

Burkhardt, P.K., Beyer P., Wunn Y., Kioti A., Amstrong G.A., Schledz M., von Liting I. and Potrykus I. 1997. *Plant J.*, **11:** 1071-1078.

Coen, E.S. and Meyerowitz, E.M. 1991. Genetic interactions controlling flower development. *Nature.* **353** No. 6339, pp. 31-37.

Guy, P. 1998. *Biodiversity.* Bernard Foundation, pp. 51-58.

Meyerowitz, E.M. 1998. Genetic and molecular mechanisms of pattern formation in *Arabidopsis* flower development. *J. Plant. Res.*, **111:** 233-242.

Tempe, J. and Schell, J. Genetic manipulation of plants. *La Recherche*, no. 188, May 1987, pp. 696-709.

Index